EXPERIMENTATION

An Introduction to Measurement Theory and Experiment Design

Third Edition

D. C. Baird

Royal Military College
Kingston, Ontario

Prentice-Hall, Inc.
Englewood Cliffs, New Jersey 07632

Library of Congress Cataloging-in-Publication Data

Baird, D. C. (David Carr)
  (Experimentation : an introduction to measurement theory and
experiment design / D. C. Baird. — 3rd ed.
    p. cm.
  Includes bibliographical references and index.
  ISBN 0-13-303298-1
  1. Physical measurements. I. Title.
  QC39.B17   1995
  507'.24—dc 20   94–13384
                                 CIP

Acquisition Editor: Ray Henderson
Assistant Acquisition Editor: Wendy Rivers
Editorial Assistant: Pam Holland-Moritz
Production Editors: Rose Kernan and Fred Dahl
Copy Editor: Rose Kernan
Production Coordinator: Trudy Pisciotti

ISBN 0-13-303298-1

Prentice-Hall International (UK) Limited, *London*
Prentice-Hall of Australia Pty. Limited, *Sydney*
Prentice-Hall Canada Inc., *Toronto*
Prentice-Hall Hispanoamericana, S. A., *Mexico*
Prentice-Hall of India Private Limited, *New Delhi*
Prentice-Hall of Japan, Inc., *Tokyo*
Simon & Schuster Asia Pte. Ltd., *Singapore*
Editora Prentice-Hall do Brasil, Ltda., *Rio De Janeiro*

TO MARGARET

Contents

Chapter 4: Scientific Thinking and Experimenting, 57

Chapter 5: Experiment Design, 84

Chapter 6: Experiment Evaluation, 113

Preface

The first edition of this book was written with the conviction that, regardless of the chosen objectives for an introductory physics laboratory, the basic principles of experimenting should not be neglected and could in fact become the principal topic. Introductory laboratories in physics are particularly suited to this purpose since the systems and theories found there are usually simple enough that the basic characteristics of measurement and experimenting can easily be made visible and understandable Such an approach to physics laboratory work can, therefore, be beneficial for a wide range of students, not only those who will proceed to professional work in physics.

That purpose on which the 1962 edition was based seems still to be valid. Many changes have taken place in the practice of experimenting, partly through the introduction of new instrumentation, but mostly because of the revolutionary impact of computing. Not only can we easily attain a level of post-experiment data analysis that would have been completely impracticable 30 years ago, but the possibilities for the conduct of the experiment itself have been enormously expanded by the availability of on-line data analysis and computer-based control of the apparatus.

Revolutionary though such changes have been in the actual conduct of experiments there has, nevertheless, been little or no change in the basic principles underlying the experimenting, and training in these basic principles is still required. Indeed, emphasis on these basic principles may be even more necessary today than it was 30 years ago on account of the present-day possibility than an experimenter can be completely insulated from the phenomena under study by an almost impenetrable barrier of data processing equipment and procedures. Under these circumstances, wholly invisible defects can produce final results with little or no meaning. Unless we have complete and clear understanding of all phases of our experiment and data analysis, we turn over our experiment wholly to the computer at our peril.

The plan of the present edition is largely the same as in that first edition but the text has been almost completely rewritten. Chapter 1 gives an outline of an approach to introductory physics laboratory work that emphasizes the basic nature of experimenting. Chapters 2, 3, and 4 provide the information on measurement, statistics, and scientific procedures that is needed to understand the basic principles of experiment design. Chapter 5 treats in a step-by-step way the practical requirements in designing an experiment, and Chapter 6 provides the corresponding procedures for evaluating the results of the experiment after the measurements have been made. At the end of the main text, Chapter 7 contains some suggestions for writing laboratory reports. The main addition to the present edition is material on computer processing of experimental observations.

The appendices contain material which, although desirable in itself, would have interrupted the development within the main text. This includes mathematical derivation of some of the equations quoted in the main text. In addition, a sample experiment is described in extensive detail, starting at the beginning of the experiment design, continuing through the conduct of the actual experiment and the evaluation of results, and ending with the final report.

The material in the text has been based on many years of teaching in our First Year Physics Laboratory, and I am grateful to the generations of students whose sometimes painful experience with it provided the opportunity for continued refinement. I wish to express my appreciation, too, to Mr. Peter Snell for valuable discussions and review of the text.

D.C. Baird

1

Approach
to Laboratory Work

This book is intended for use in introductory physics laboratories. It was written, however, in the hope that it will serve a much wider purpose. It provides an introduction to the study of experimenting in general, irrespective of the area in which the experimenting is carried out. Some students in an introductory physics laboratory may pursue careers in physics research, and it is hoped that the book will serve as a suitable introduction to their continued studies. Many others will pursue careers in completely different areas, perhaps in other sciences, or perhaps in areas outside science altogether. Whatever the need, the introductory physics laboratory can, if suitably oriented, provide a useful introduction to the fundamental principles that underlie experimenting of any kind. For our purposes, the word *experimentation* has a very broad definition. By it we mean the whole process of identifying a portion of the world around us, obtaining information from it, and interpreting that information. This definition covers a wide range of activities—all the way from the traditional picture of a biologist in a white coat splicing DNA molecules to a manufacturer taking a poll to determine individual preferences in toothpaste. This book is intended to meet the needs of all who are either engaged themselves in any kind of study of the world around us, or who must form a judgment on scientific statements made by others.

1–1 NATURE OF SCIENTIFIC KNOWLEDGE

It is natural to ask why everyone, not just practicing scientists, should be familiar with the processes by which we gain knowledge of the world around us. The answer is found in the extensive part experimenting plays in our lives, whether we are aware of it or not. Even if we are not practicing scientists ourselves, almost all of us in our daily lives are faced with the requirement to use or to pass judgment on experimental information offered by others. For example, our professional work may require us to make a choice between competitive bids on equipment having certain specifications, or as members of the general public, we may be called on to form opinions on such issues as the possible health hazards of nuclear power plants, the safety of food additives, the impact of acid rain on the environment, or the influence of national monetary policy on unemployment. All these issues require us to be familiar with the nature of the processes of scientific experiments and to make decisions that are based on our own, appropriately skeptical assessment of the reliability of the experimental information.

To do this we must first become knowledgeable about the nature of measurement itself. In specific, we must be clearly aware that measured values cannot be exact. What we call the **uncertainty** in the measurement can arise either from limitations of the instruments or from statistical fluctuation in the quantity being measured. Whatever the origin of the uncertainty, we must be aware of its existence and know how to estimate it. Only then can we know how much confidence to place in the measured value.

Once the nature of measurement has been settled, a second, equally significant requirement remains. Despite all the best efforts of science educators worldwide, much misunderstanding still remains concerning the basic nature of scientific statement. The misunderstanding commonly centers on the question of the reliability or authority of the statements. Attitudes vary all the way from unquestioning faith that some point has been "scientifically proved," and so must be infallibly correct, to complete scepticism that all science is "just theories," and so can be confidently disregarded.

As can easily be expected, neither extreme position is valid, and public life will be better served if we are all able to take the scientific or technical statements we hear and place them appropriately along the scale of credibility. To be able to do so, one vital point must be appreciated even before we turn our attention to the mechanical details of the processes by which the information is obtained. It is a point that is frequently neglected in public statements about scientific or technical affairs, and yet it is indispensable for their proper comprehension.

The point is this. There exists an all-important distinction between the portion of the real world that is under discussion (we refer to it as the **system under study**) and the ideas and concepts (generally referred to as the **model**) that we invent in our heads as a consequence of observing the system. There is usually little difficulty in comprehending the nature and status of the first of these, the measurements. The second, however, needs a little elaboration. Our purpose in inventing ideas is to represent the observed properties of the system in a kind of shorthand way so that we can talk to each other about the system conveniently, easily, and with a common basis of understanding. For example, if we were the first people on earth exploring our environment for the first time, we might notice each day a certain type of tree in our wanderings. But instead of reporting today's sighting and yesterday's sighting separately, and so on for the preceding month, as if each observed event were unrelated, it would be much more convenient to invent, using a certain set of described properties, the abstract concept "banana." Such a concept would allow us to plan together for tomorrow's dinner much more expeditiously and with better economy of communication than would be the case without it. Beyond such simple examples, the use of models is widespread, significant, and sophisticated.

As we pursue our daily lives, it is easy to forget that many of the items that appear in routine communication concern concepts and ideas and are not genuine statements about the real world. Frequently, the distinction is completely unimportant, and we can get away with careless use of language. Occasionally, however, the distinction is vital, and serious error can result from inattention to it.

The danger arises because the two aspects of our knowledge of the external world have completely different character. On the one hand, observations that we make on our system belong to the real world and can (subject to the necessary presence of uncertainty that is discussed in Chapter 2) have the status of genuine incontrovertibility. As an example, no thinking person would question the assertion that we can measure the width of the Atlantic Ocean to be greater than the length of our living room. With varying degrees of confidence, therefore, all statements about *observations* of the real world have the potential of being immune against refutation, and this can mislead us by falsely reinforcing the notion that *all* scientific statements contain incontrovertible truth about the universe.

Unlike statements about measurements, however, statements about our ideas and concepts have no claim whatsoever as absolute knowledge about the universe. They are nothing more than ideas invented in our heads, and even if they are ideas that have been very carefully chosen to represent the properties of the system as closely as possible, they remain nothing more than just that: ideas in our heads. Not only must they remain provisional, subject to improvement or replacement if someone comes along with a better idea, but it is impossible for them to qualify, as can observations, as certain knowledge of the external world.

Failure to comprehend the complementary roles of observation and concept in science is very common in public debates about scientific matters and is the source of much confusion. For example, a famous economist was recently heard to marvel publicly that a drop in interest rates had failed to provide the desired stimulus to the economy. He neglected to mention that the hoped-for relationship between interest rates and economic activity belonged to his *model*; the real economy, of course, has ideas of its own.

All statements that come from scientific study of a system fall into one or other of the categories that have been mentioned. They may be statements about *observations* of a system, they may be statements about *models*, or they may be statements about the *relationship* between a system and a model. If we keep these possibilities clearly in mind as we listen to scientific statements and analyse them into these categories, we shall have gone far toward forming accurate judgements.

As far as our own statements on scientific matters are concerned, we should accept the responsibility to be precise about the language we use. We still hear pronouncements from famous scientists that a "correct theory" for something has been found. They may themselves have clear understanding about what they mean, and they do not mislead those of us who know how to decode such conventional language. But the potential for misunderstanding in nonscientific circles is too serious, and all of us who make scientific statements should watch our language very carefully.

1–2 ROLE OF THE PHYSICS LABORATORY

But what does all this have to do with possible purposes for the introductory physics laboratory? Physics teaching laboratories have played such a familiar role that it is natural to wonder how the normal laboratory with its usual experiments can be used to provide an introduction to experimenting in general. The answer lies not so much in the experiments themselves as in the attitude with which we approach them. This will become clear as our studies of experimental methods proceed, but it may be helpful at this point to illustrate the suggestion by anticipating in outline the arguments that are discussed in more detail later. We have already mentioned that our studies on experimental methods will be clarified if we regard the piece of the universe under study as a **system**. By a *system* we mean, in general, any isolated, defined entity that functions in a specific manner. We assume that we can influence or control the system, and we refer to the methods we have available to do this as **inputs**. We also assume that the system performs some identifiable function or functions, and we refer to these as **outputs**. The various examples that follow will make clear the use of the terminology. An economist, for example, may view the economy of a country as a system with an extensive set of inputs and a correspondingly varied set of outputs. The system itself includes the whole productive capacity for goods and services, transportation facilities, supply of raw materials, inhabitants, opportunities for foreign trade, weather, and many other things. The inputs are those things that can be controlled by us—the money supply, tax rates, government spending, tariffs on imports, and so on. The outputs are those things that we *cannot* control directly; their magnitudes are determined by the system, not by us. Outputs of an economic system include the gross national product, unemployment rate, inflation rate, external trade balance, and the like. It would be very comforting and convenient if we could secure the desired values of these outputs by simple manipulation, but we cannot. No matter how desirable it may be, we cannot instruct the country's gross national product or unemployment rate to have a certain value; we are restricted to controlling the inputs. Even there we have problems. In a system as complex as a national economy, the linkages between the inputs and the outputs are tangled and indirect. A change in one input variable will likely have an effect on a number of output variables, instead of solely on the single output in which we may be interested. For example, an attempt to increase the gross national product of a country by reducing taxation rates will possibly be at least partially successful, but the simultaneous effects on other outputs may be equally prominent and not nearly as desirable—for example, a possible increase in the rate of inflation. The methods available for handling such situations are sophisticated, but given the complexity of the system, the level of success achieved by the politicians and economists shows that substantial room for improvement still remains.

There are other systems that, although still complex, are simple enough for us to control them reasonably successfully. Consider, for example, an electrical power generating and distribution system. It, too, is an enormous and complex system. It has many inputs, such as the number of generators that are started up, the routing of the power through the transmission lines, the salaries paid to the staff, the hours of work on the shifts, the price charged to the customers per kilowatt-hour, and so forth. It has obvious outputs, such as the power delivered to various districts, and also other, less tangible outputs, such as the overall efficiency of the system and the reliability of service. These are quantities that cannot be controlled directly by the management; the system decides their values, and so they must be counted as outputs of the system.

How does all this refer to the introductory physics laboratory? If we are to prepare people to enter a scientifically literate population, would it not be better to tackle the important problems right away and start deciding whether the mercury content of fish makes it safe to eat or whether our use of fossil fuels is contributing to global warming? The trouble is that these are extremely difficult problems. Evidence is hard to obtain and its interpretation is usually uncertain; even the experts themselves disagree, often vigorously and publicly. It is almost impossible to make a significant contribution to the solution of such complex problems without first developing our skills through using simpler situations.

To see how this can be done, let us think about some of these simpler systems. An automobile engine is a system that is simple in comparison with any of the earlier examples. The system includes only the engine, fuel supply, mounting, surrounding atmosphere, and so on. The inputs may be the obvious controls such as fuel supply, fuel–air ratio, and ignition timing (even though some of our direct control may have been usurped by the computerized systems now found in automobiles, these quantities remain as inputs to the engine itself). The output as always are the factors whose values are set by the system—for example, the number of rpm, the amount of heat produced, the efficiency of energy conversion, and the composition of the exhaust gases. This is still a somewhat complex system, but we can begin to see that relatively simple relationships between inputs and outputs can exist. For example, the input–output relation between accelerator setting and rpm for a gasoline engine is sufficiently direct and predictable for most of us to invoke it daily. Notice, however, that the system is still sufficiently complex so that the effect of that one input is still not restricted to the single output in which we are interested—rpm. Other outputs such as heat produced, composition of exhaust gas, and efficiency are also affected by the accelerator setting, even if we are generally prepared to ignore the connection.

In the example of the automobile engine, we are beginning to reach the stage at which the system is simple enough for us to start identifying basic principles of experimenting. Let us go one stage further and consider the example of a simple pendulum. It, too, is a system but one that includes very little other than the string, the mass at the end of the string, the supporting stand, and the surrounding air. Furthermore, it has only two immediately obvious inputs—the length of the string and the initial conditions according to which the motion is started. The outputs, too, are few in number. Apart from small, secondary effects, they include only the frequency of vibration and the amplitude of oscillation. Last, the connection between the inputs and outputs is relatively direct and reproducible. Altering the length of the pendulum's string offers few surprises when we measure the frequency of vibration. Here is a system that is simple enough to allow the basic principles of experimenting to be clearly visible. If we use it to develop expertise in studying systems and evaluating their outputs, we shall acquire the competence to tackle later the more important but more complex problems.

This gives us the key to at least some constructive uses of the introductory physics laboratory. There *is* real point in working with a pendulum but only if we view it properly. If we look at it as "just a pendulum," our only reaction will be total boredom. If, however, we view it as a system, just like a supermarket, an electrical power system, or the national economy, but differing from these only in that it is simple enough for us to understand it relatively well, our battered old pendulum supplies excellent simulation of the problems of the real world.

The introductory physics laboratory, therefore, can offer us the opportunity to practice on simple systems and develop the expertise that we shall require later in the real world with its important but complicated systems. We must be careful, however, about the ways we practice on these simple systems. We shall derive only very limited benefit if, for example, we restrict ourselves to sets of instructions that tell us exactly how to do particular experiments. The range of experimental situations in society is enormous. In some areas, random fluctuation dominates, as in the biological sciences; in others, measurement may be precise, as in astronomy, but control over the subject matter is limited. If we are to learn to function independently within this enormous range, it is necessary to identify *general* principles of experimenting that can be applied later to any future subject matter or type of experimenting. The remainder of this text is concerned with some of those principles, and we assume henceforth that laboratory experiments will be regarded as exercises to illustrate the principles.

It may now be obvious that many of the traditional procedures in introductory laboratories are inappropriate for our purpose. For example, we must avoid thinking of an experiment as a procedure to reproduce some "correct" answer, deviation from which makes us "wrong." Instead, we should simply assess the properties of our particular system dispassionately and take the results as they come. Also, there is no point in seeking some "procedure" to follow: that is nothing more than asking someone else to tell us how to do the experiment. In real life we rarely find someone waiting to tell us what to do or what our result should be; our usefulness depends on our ability to make our *own* decisions about how to handle a situation. It takes a great deal of practice and experience to develop confidence in our own decisions about experiment procedures, and the introductory laboratory is not too early to start. We therefore place a great deal of emphasis on experiment planning, for this is the stage at which much of the skill in experimenting is needed. It is important to avoid the temptation to regard preliminary planning as a waste of time or a distraction from the supposedly more important task of making the measurements. Time must be explicitly set aside for adequate analysis and planning of the experiment before a start is made on the actual measuring process.

It is also necessary to learn to work within the framework of the apparatus available. All professional experimenting is subject to limits on resources, and much of the skill in experimenting lies in optimizing the yield from these resources. Restrictions on time, too, merely simulate the circumstances of most actual experimenting. The apparatus itself is not always good enough, but this should not be regarded as a defect but as a challenge, for this aspect as before accurately simulates real life. Good evaluation of experimental results always requires us to separate the grain of useful measurement from the chaff of error, uncertainty, and mistaken interpretation. Experimenters must learn to identify sources of error or uncertainty for themselves and, if possible, eliminate them or correct for them. Even with the greatest care there is always an irreducible residuum of uncertainty, and it is experimenters' responsibility to evaluate it accurately. The ability to cope with such requirements can be acquired only by actual contact with realistic working conditions. It is a common injustice to students in introductory physics laboratories to provide apparatus that is too carefully adjusted or manipulated, which can give students the impression that experiments always give the "right" answer. This is unfortunate, because the foundation of future expertise lies in learning how to respond constructively to the limitations of experiments.

The most fruitful use of laboratory time results when the experiments are accepted as problems that we must solve on our own. Certainly, errors in judgment will be made, but students can learn more effectively when they see the consequences of their decisions through direct, personal experience than when they rigidly follow some established "correct" procedure. What is learned from an experiment is more important than the production of some supposedly "good" result. This is not to say that they should be complacently indifferent to the outcome of the experiment. Development of experimenting skills comes about only if the challenge of obtaining the best possible result in every experiment is taken seriously.

The writing of laboratory reports should be tackled in the same constructive spirit. In professional life there is very little point in spending time and trouble on an experiment unless we can adequately convey the outcome to others. We have an obligation to our readers to express ourselves as lucidly, if not elegantly, as possible. It is wrong to regard this as solely the responsibility of our local Departments of English. Report writing in the introductory science laboratory is an opportunity for exercise in descriptive composition. Report writing that degenerates into a mere indication that the experiment has been performed is little more than a waste of time and a loss of opportunity for necessary practice. Report writing at the level suggested here is almost pointless without adequate review and criticism. Opportunities for improvement become much more obvious in hindsight, and such detailed review should be regarded as an indispensable part of the work in a teaching laboratory.

2

Measurement
and Uncertainty

2-1 BASIC NATURE OF MEASURING PROCESS

Measurement is the process of quantifying experience of the external world. The nineteenth-century Scottish scientist, Lord Kelvin, once said that "when you can measure what you are speaking about and express it in numbers, you know something about it; but, when you cannot measure it, when you cannot express it in numbers, your knowledge is of a meager and unsatisfactory kind; it may be the beginning of knowledge, but you have scarcely in your thoughts advanced to the stage of science." Although this may be a slight overstatement, it remains true that measurements constitute one of the indispensable ingredients of experimenting. We cannot reach a satisfactory level of competence in experimenting without knowledge of the nature of measurement and the significance of statements about measurements.

It is obvious that the quantifying process (almost invariably involves comparison with some reference quantity (how many paces wide is my backyard?). It is equally obvious that the good order of society requires extensive agreement about the choice of these reference quantities. Such measurement standards, defined by legislation and subject to international agreement, are extensive and important. No one seriously interested in measurement can ignore defining and realizing such standards in his or her area of work. A discussion of this important topic here would distract us from our chief concern, the actual process of measuring, so we leave the topic of standards without further mention (except reference to the texts listed in the Bibliography), and take up the study of measuring processes.

We start at the most basic level with an apparently simple measurement and use it to find out what kind of process is involved and what kind of statement can be made about its outcome. If I give someone my three-ring binder containing this text with the request to measure its length with a meter stick, the answer is absolutely invariable: The length of the notebook is 29.5 cm. But that answer must make us wonder: Are we really being asked to believe that the length of the book is exactly 29.50000000···cm? Surely not; such a claim is clearly beyond the bounds of credibility. So how are we to interpret the answer? A moment's thought in the presence of the notebook and a meter stick makes us realize that, far from determining the "right" or "exact" value, the only thing we can realistically do is to approach the edge of the notebook along the scale, saying to ourselves as we go: Am I sure the answer lies below 30 cm? Below 29.9 cm? Below 29.8 cm? The answer to each of these questions will undoubtedly be: Yes. As we progress along the scale, however, we eventually reach a point at which we can no longer give the same confident reply. At that point we must stop and we identify one end of an interval that will become our measured value. In a similar way we can approach the edge of the notebook from below, asking ourselves at each stage: Am I sure that the answer lies above 29.0 cm? Above 29.1 cm? And so on. Once again, we reach a value at which we must stop, because we can no longer say with confidence that the answer lies above it. By the combination of these two processes we identify an interval along the scale. It is the smallest interval that, as far as we can be certain, does contain the desired value; within the interval, however, we do not know where the answer lies. Such is the only realistic outcome of a measuring process. We cannot look for exact answers; we must be content with measured values that take the form of intervals. Not only does this example illustrate the essential nature of a measuring process, it also provides guidance for actually making measurements. The process of approaching from each side separately the value we seek is a reminder of the necessity of stating the result as an interval, and it also makes it easier to identify the edges of that interval.

The final outcome of our discussion is most important. As we make measurements and report the results, we must constantly keep in mind this fundamental and essential point—measurements are not exact, single numbers but consist of *intervals*, within which we are confident that our desired value lies. The act of measuring requires us to determine both the location and width of this interval, and we do it by the careful exercise of visual judgment every time we make a measurement. There are no simple rules for determining the size of the interval; that depends on many factors in the measuring process. The type of measurement, the fineness of the scale, our visual acuity, the lighting conditions—all play a part in determining the width of the measurement interval. The width, therefore, must be determined explicitly each time a measurement is made. For example, it is a common error to believe that, when a measurement is made using a divided scale, the "reading error" is automatically one-half of the finest scale division. This is an erroneous oversimplification. A finely divided scale used to measure an object with ill-defined edges can give a measurement interval as large as several of the finest scale divisions. A well-defined object and good viewing conditions, on the other hand, may permit the identification of a measurement interval well within the finest scale division. Every situation must be assessed individually.

2–2 DIGITAL DISPLAY AND ROUNDING OFF

Other aspects may confuse the issue. Consider, for example, a piece of equipment that gives a digital readout. If a digital voltmeter shows that a certain potential difference is 15.4 V, does that readout imply that the value is 15.40000 ... exactly? Clearly not, but what does it mean? That depends on circumstances. If the instrument is made in such a way that it reads 15.4 V because the actual value is closer to 15.4 than it is to 15.3 or 15.5, then the meaning is: This reading lies between 15.35 V and 15.45 V. On the other hand, a digital clock may be made in such a way that it changes its indication from 09.00 to 09.01 at the time of 09.01. When we see it reading 09.00, we know that the time lies between 09.00 and 09.01, a slightly different interpretation from that appropriate to the digital voltmeter. Again, each situation must be judged by itself.

These two examples of digital display illustrate a more general concept—the inaccuracy inherent in the process of rounding off. Even without the uncertainty that arises from limited ability to make measurements, the mere statement of a numerical quantity can contain uncertainty. Consider the statement

$$\pi = 3.14$$

We all know that this is not so because we can remember some of the numbers that follow: 3.14159 ... and so on. So what can we mean by quoting π as 3.14? Presumably, we mean only that π has a value closer to 3.14 than it does to 3.13 or 3.15. Our statement, therefore, can be translated to read—π lies between 3.135 and 3.145. This range of possibility represents what is sometimes known as a *rounding-off error*. The effect of such errors can be small and unimportant, or they can become significant. In a long calculation, there is a chance that rounding-off errors either can accumulate, or they can be important in other ways. For example, the calculation may require us to find the difference between two large calculated values. If these two calculated values are close together, the result that we need may be greatly affected by premature rounding off. Because calculators make accurate calculation so easy, it is always advisable to carry the calculation through more figures than one might initially think would be necessary. We can always do appropriate rounding off at the end of the calculation.

A similar rounding-off error can appear in statements about measurement. We sometimes hear that someone has made a measurement on a scale that was "read to the nearest millimeter" or some such phrase. This is not a very good way of reporting a measurement because it obscures the actual value of the measurement interval. When we encounter such statements, we can only assume that the quoted scale division represents some kind of minimum value for the size of the measurement interval.

2–3 ABSOLUTE AND RELATIVE UNCERTAINTY

Whatever the means by which we make a measurement, the final outcome should be the interval that best represents the range inside which the desired value lies. In the example we used first, the experimenter might be able to state with confidence nothing more than that, as a consequence of a certain measuring process, the length of the notebook lay between 29.4 and 29.6 cm. This is a perfectly satisfactory and accepted way to quote the result. Although such a statement corresponds with the reality of the measuring process, it is frequently desirable to *rephrase* the quoted value. Take the interval 29.4 to 29.6 cm and *rename* it 29.5 ± 0.1 cm. Although this expression is no more than a renamed statement of the original interval, the new form does offer certain advantages. It gives us a central value, 29.5 cm, which can be used in further calculations. It also gives us a value, ±0.1 cm, that we call the **uncertainty** of the measurement. First, the magnitude of the uncertainty enables us to judge the quality of the measuring process. Second, we can use this numerical measure of the uncertainty in continued calculations on uncertainties. One disadvantage of this mode of expression is the return to a central value, 29.5 cm. Unless we remember clearly that only the complete quantity, 29.5 ± 0.1 cm, serves as an adequate statement of the answer, we may become sloppy in making and reporting measurements and may forget the essential presence of the uncertainty. It should be an invariable practice to associate an uncertainty value with a reading, both at the time of the measurement and subsequently, whenever the value is quoted or used in further calculation.

Because the figure ±0.1 cm represents the actual amount, or range, by which the reading of 29.5 cm is uncertain, it is often called the **absolute uncertainty** of the measurement, and we consistently use this terminology. For the purpose of perceiving the *significance* of the uncertainty, however, it is frequently convenient to extend the definition of uncertainty. How significant is an uncertainty of ±0.1 cm? When we measure the length of a notebook, it is significant to a certain extent. When we measure the distance between two cities, an uncertainty of ±0.1 cm surely is completely insignificant. At the other end of the scale, however, if we measure the size of a microscopic bacterium, an uncertainty of ±0.1 cm clearly makes the measurement meaningless. Obviously, the significance of a particular uncertainty value depends on the magnitude of the measurement itself. For this reason, it is frequently desirable to compare an uncertainty figure with the actual value of the measurement. For this purpose, we define a quantity called the **relative uncertainty** of the measurement. It is defined by

$$\text{Relative Uncertainty} = \frac{\text{Absolute Uncertainty}}{\text{Measured Value}}$$

In the case of our example

$$\text{Relative Uncertainty} = \frac{\pm 0.1}{29.5} = \pm 0.003$$

This relative uncertainty is often quoted as a percentage, so that in the present case the relative uncertainty is ±0.3%. Such a quantity gives us a much better feeling for the quality of the measurement, and we often call it the **precision** of the measurement. The absolute uncertainty has the same dimensions and units as the basic measurement (29.5 cm is uncertain by 0.1 *cm*), whereas the relative uncertainty, being a ratio, has neither dimensions nor units and is a pure number.

2–4 SYSTEMATIC ERRORS

The kind of uncertainty that we have been considering arises from naturally occurring limitations in the measuring process. A different type of error can appear when something affects all the measurements of a series in an equal or a consistent way. For example, a voltmeter or a micrometer caliper can have a zero error, a wooden meter stick may have shrunk, a stopwatch may be running fast or slow, and so on. These errors are termed **systematic errors**, a subclass of which are **calibration errors**. Because such systematic errors may not be immediately visible as one

makes a measurement, it is necessary to be vigilant and remember at all times the possibility of their presence. Instrument zeroes, for example, should automatically be checked every time an instrument is used. Although it may be less easy to check calibration, the accuracy of electrical meters, timing devices, thermometers, and other such instruments should not be taken for granted and should be checked whenever possible. Also, the presence on an instrument of a precise-looking, digital readout with four or five supposedly significant figures should not be taken as proof of precision and freedom from systematic error. Most of a batch of electronic timers that our laboratory once acquired for laboratory teaching, which could supposedly measure time intervals with millisecond accuracy, turned out to have calibration errors as large as 14%. Do not be deceived; view all measuring instruments with suspicion and check instrument calibration whenever possible.

2–5   UNCERTAINTY IN CALCULATED QUANTITIES

The preceding sections have been concerned solely with the concept of uncertainty in a single measurement. It is rare, however, that a single measurement ends the process. Almost invariably the result we desire is a combination of two or more measured quantities or is at least a calculated function of a single measurement. We might wish, for example, to calculate the cross-sectional area of a cylinder from a measurement of its diameter, or to calculate its volume from measurements of both diameter and length. The various measurements will sometimes be of different types, as in a calculation of g from values of the length and period of a pendulum. In all such cases the presence of uncertainty in the basic measurements obviously entails the presence of uncertainty in the final computed value. It is this final uncertainty that we now wish to calculate. For the purposes of this section we assume that our uncertainties have the character of ranges or intervals within which we are "almost certain" that our answer lies. For the computed values we calculate intervals within which we again wish to be "almost certain" that our answer lies. That means that we must do our calculation for the "worst case" of combined uncertainties in which the deviations in the various measured quantities happened to occur in such directions as to reinforce each other. This is perhaps a pessimistic assumption, and we see later in Chapter 3 how the probabilities that are associated with various error combinations enable us to make a more realistic and less pessimistic estimate. For the moment, assume that we wish to calculate from the uncertainties in the primary values the maximum range of possibility for the computed answer.

2–6 UNCERTAINTY IN FUNCTIONS OF ONE VARIABLE ONLY

Consider a measured quantity x_0 with an uncertainty $\pm \delta x$ (where we are using finite differences such as δx to represent absolute uncertainties in the corresponding variable x, etc.), and consider a computed result z to be some function of the variable x. Let

$$z = f(x)$$

The function f enables us to calculate the required value z_0 from a measured value x_0. Moreover, the possibility that x can range from $x_0 - \delta x$ to $x_0 + \delta x$ implies a range of possible values of z that range from $z_0 - \delta z$ to $z_0 + \delta z$, where δz is the value of the absolute uncertainty in z. We now wish to calculate the value of δz. The situation is illustrated graphically in Figure 2–1, in which, for a given $f(x)$, we can see how the measured value x_0 gives rise to the computed result z_0, and how the range $\pm \delta x$ about x_0 produces a corresponding range $\pm \delta z$ about z_0.

Before considering general methods of evaluating δz it is instructive to see how finite perturbations are propagated in simple functions. Consider, for example, the function

$$z = x^2$$

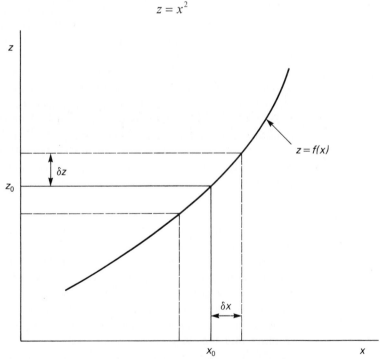

Figure 2-1 Propagation of uncertainty from one variable to another.

If x can range between $x_0 - \delta x$ and $x_0 + \delta x$, then z can range between $z_0 - \delta z$ and $z_0 + \delta z$, where

$$z_0 \pm \delta z = (x_0 \pm \delta x)^2$$
$$= x_0^2 \pm 2x_0 \delta x + (\delta x)^2$$

We can ignore $(\delta x)^2$, since δx is assumed to be small in comparison with x_0, and equate z_0 to x_0^2, giving for the value of δz

$$\delta z = 2x_0 \delta x$$

This can more conveniently be expressed in terms of the relative uncertainty $\delta z / z_0$

$$\frac{\delta z}{z_0} = \frac{2x_0 \delta x}{x_0^2} = 2\frac{\delta x}{x_0}$$

Thus, the relative uncertainty of the computed result is twice that of the initial measurement.

Although it is helpful to bear in mind the nature of propagated uncertainty, as illustrated by the use of finite differences, considerable simplification of the formulation can be achieved using differential calculus.

2–7 GENERAL METHOD FOR UNCERTAINTY IN FUNCTIONS OF A SINGLE VARIABLE

The finite differences δz and δx that were used in the preceding section could be regarded as components of the derivative dz / dx. We can therefore obtain our value of δz by first using standard techniques to obtain dz / dx in the form

$$\frac{dz}{dx} = \frac{d(f(x))}{dx}$$

and then writing

$$\delta z = \frac{d(f(x))}{dx} \delta x \tag{2-1}$$

This is a relatively simple procedure, and it works well in cases for which the elementary, finite-difference approach would lead to algebraic complexity. If, for example we have to deal with the function

$$z = \frac{x}{(x^2 + 1)}$$

then

$$\frac{dz}{dx} = \frac{x^2 + 1 - x \times 2x}{(x^2 + 1)^2}$$

$$= \frac{1 - x^2}{(1 + x^2)^2}$$

and

$$\delta z = \frac{1 - x^2}{(1 + x^2)^2} \delta x$$

This calculation would have been very awkward if any other approach had been used. Furthermore, it gives δz generally as a function of x and δx; any particular desired value can be obtained by setting $x = x_0$. We now use this technique to evaluate uncertainties for some common functions.

Powers

Consider the function

$$z = x^n$$

Differentiating and replacing the derivative by finite differences gives us

$$\frac{dz}{dx} = nx^{n-1}$$

$$\delta z = nx^{n-1} \delta x$$

The significance of this result becomes a little more obvious when expressed in terms of the relative uncertainty. Thus,

$$\frac{\delta z}{z} = n \frac{\delta x}{x}$$

Thus, when evaluating powers of a measured quantity, we must compute the uncertainty in the final answer using the *relative uncertainty* of the measured quantity. The relative uncertainty in the final answer is the relative uncertainty of the basic quantity multiplied by the power involved. This method is valid for either powers or roots, so that precision diminishes as a quantity is raised to powers and improves on taking roots. This situation must be carefully watched in an experiment in which powers are involved. The higher the power, the greater is the need for good initial precision.

Trigonometric Functions

We give only one example, because all the others can be treated in similar ways. Consider

$$z = \sin x$$

Here

$$\frac{dz}{dx} = \cos x$$

and

$$\delta z = (\cos x)\delta x$$

This is one case where the elementary method of inserting $x_0 \pm \delta x$ in the function shows the result more clearly. We obtain

$$z_0 \pm \delta z = \sin(x_0 \pm \delta x)$$
$$= \sin x_0 \cos \delta x \pm \cos x_0 \sin \delta x$$

Since $z_0 = \sin x_0$ and $\cos \delta x \approx 1$, this becomes

$$\delta z = \cos x_0 \sin \delta x$$

This result makes it clear that the δx in the original expression was really an approximate form of $\sin \delta x$. Only in the case of very large uncertainty would this difference be significant, but it is best to be aware of the situation. For one thing, it allows us to understand that, when we use that original expression, we must express the uncertainty in angle, δx, in *radian* measure, because only then can we replace $\sin \delta x$ by δx itself. Such uncertainty calculations using trigonometric functions normally have straightforward application when dealing with apparatus such as spectrometers.

Logarithmic and Exponential Functions

Consider the function

$$z = \log x$$

Here

$$\frac{dz}{dx} = \frac{1}{x}$$

and

$$\delta z = \frac{1}{x}\delta x$$

and the relative uncertainty can be calculated as usual.

If

$$z = e^x$$

$$\frac{dz}{dx} = e^x$$

and

$$\delta z = e^x \delta x$$

This is an important case, because exponential functions occur frequently in science and engineering. These functions can become very sensitive to the exponent when it takes values much over unity, and the uncertainty δz may become very large.

As stated earlier, the method can be easily applied to any function not listed above by evaluating the appropriate derivative and using Equation (2–1).

2–8 UNCERTAINTY IN FUNCTIONS OF TWO OR MORE VARIABLES

If the result is to be computed from two or more measured quantities, x, y, and so on, the uncertainty in the result can, as was mentioned in Section 2–5, be regarded in two ways. We could be as pessimistic as possible and suppose that the actual deviations of x and y happen to combine in such a way as to drive the value of z as far as possible from the central value. In this way, we would calculate the value for δz that gives the extreme width of the range of possible z values. On the other hand, we could argue that it is more probable for the uncertainties in the basic measurements to combine in a less extreme way, some making positive contributions to δz and some negative, so that the resulting δz would be smaller than for the pessimistic assumption. This argument is valid, and later we deal with the question of *probable* uncertainty in computed quantities. For the moment, however, we calculate the value of δz that represents the widest range of possibility for z. Such an approach, if pessimistic, is certainly safe, because, if δx, δy, and so forth, represent limits within which we are "almost certain" the actual values lie, then the calculated δz will give those limits within which we are equally certain that the actual value of z lies.

The most instructive initial approach uses the elementary substitution method, and we use this for the first two functions.

Sum of Two or More Variables

Consider the function of two variables

$$z = x + y$$

The uncertainty in z will be obtained from

$$z_0 \pm \delta z = (x_0 \pm \delta x) + (y_0 \pm \delta y)$$

and the maximum value of δz is obtained by choosing similar signs throughout the right-hand side of the expression. Thus,

$$\delta z = \delta x + \delta y$$

As might be expected, the uncertainty in the sum is just the sum of the individual uncertainties. This can be expressed in terms of the relative uncertainty

$$\frac{\delta z}{z} = \frac{\delta x + \delta y}{x + y}$$

but no increased clarification is achieved. If the quantity z contains the sum of more than two variables, the expression for the uncertainty in z can obviously be extended as necessary.

Difference of Two Variables

Consider a quantity that must be calculated as the difference between two measured values. Let

$$z = x - y$$

As in the preceding case, δz is obtained from

$$z_0 \pm \delta z = (x_0 \pm \delta x) - (y_0 \pm \delta y)$$

Here, however, we can obtain the maximum value of δz by choosing the *negative* sign for δy, giving again,

$$\delta z = \delta x + \delta y$$

We can see from this equation that, when x_0 and y_0 are close together and the value of $x - y$ is small, the relative uncertainty can rise to very large values. This is at best an unsatisfactory situation, and the precision can be low enough to destroy the value of the measurement. The condition is particularly hazardous because it can arise unnoticed. It is perfectly obvious that, if it were possible to avoid it, no one would attempt to measure the length of my notebook by measuring the distance of each edge from a point a mile away and then subtracting the two lengths. However, a desired result can be obtained by subtraction of two measurements made separately (two thermometers, clocks, etc.), and the character of the measurement as a difference may not be strikingly obvious. Consequently, treat all measurements involving differences with the greatest caution. Clearly, the way to avoid the difficulty is to measure the difference directly, rather than obtaining it by subtraction between two measured quantities. For example, if you have an apparatus within which two points are at potentials above ground of $V_1 = 1500$ V and $V_2 = 1510$ V, respectively, and the required quantity is $V_2 - V_1$, only a voltmeter of very high quality would permit the values of V_1 and V_2 to be measured with the exactness required to achieve even 10% precision in $V_2 - V_1$. On the other hand, an ordinary 10 V table voltmeter, connected between the two points and measuring $V_2 - V_1$ directly, immediately gives the desired result with 2% or 3% precision.

2–9 GENERAL METHOD FOR UNCERTAINTY IN FUNCTIONS OF TWO OR MORE VARIABLES

The last two examples, treated by the elementary method, suggest that the differential calculus may offer considerable simplification of the treatment. It is clear that if z is a function of the two variables x and y,

$$z = f(x, y)$$

the appropriate quantity for calculating δz is the total differential dz. This is given by

$$dz = \frac{\partial f}{\partial x} dx + \frac{\partial f}{\partial y} dy \tag{2–2}$$

We treat this differential as a finite difference δz that can be calculated from the uncertainties δx and δy. Thus,

$$\delta z = \frac{\partial f}{\partial x} \delta x + \frac{\partial f}{\partial y} \delta y$$

and the derivatives $\partial f / \partial x$ and $\partial f / \partial y$ are normally evaluated for the values, x_0 and y_0, at which δz is required. We may find that, depending on the function f the sign of $\partial f / \partial x$ or $\partial f / \partial y$ turns out to be negative. In this case, using our pessimistic requirement for the maximum value of δz, we choose negative values for the appropriate δx or δy, obtaining thereby a wholly positive contribution to the sum.

Product of Two or More Variables

Suppose

$$z = xy$$

To use Equation (2–2) we need the values of $\partial z / \partial x$ and $\partial z / \partial y$. They are

$$\frac{\partial z}{\partial x} = y \quad \text{and} \quad \frac{\partial z}{\partial y} = x$$

Thus, the value of δz is given by

$$\delta z = y\delta x + x\delta y$$

The significance of this result is more clearly seen when it is converted to the relative uncertainty

$$\frac{\delta z}{z} = \frac{\delta x}{x} + \frac{\delta y}{y}$$

Thus, when the desired quantity is a product of two variables, its *relative uncertainty* is the sum of the *relative uncertainties* of the components. Notice the contrast with the result for uncertainty in the case of two added variables, where we must combine the uncertainties using the *absolute* uncertainties.

The most general case of a compound function, very commonly found in physics, involves an algebraic product or quotient that has components raised to powers. Consider the function

$$z = x^a y^b$$

where a and b may be positive or negative, integral or fractional. This formulation is greatly simplified by taking logarithms of both sides before differentiating. Thus,

$$\log z = a \log x + b \log y$$

whence, differentiating implicitly,

$$\frac{dz}{z} = a\frac{dx}{x} + b\frac{dy}{y}$$

As usual, we take the differentials to be finite differences and obtain

$$\frac{\delta z}{z} = a\frac{\delta x}{x} + b\frac{\delta y}{y}$$

If the original expression for z contains more than two variables, we can simply extend the result for δz by adding terms as appropriate. This process gives the relative uncertainty directly, which is frequently convenient. If the absolute uncertainty δz is required, it can be evaluated simply by multiplying the relative uncertainty by the computed value z_0, which is normally available. This form of implicit differentiation still offers the simplest procedure even when z itself is raised to some power. For example, if the equation reads

$$z^2 = xy$$

it is unnecessary to rewrite it

$$z = x^{1/2} y^{1/2}$$

and work from there, because, by taking logs

$$2 \log z = \log x + \log y$$

whence

$$2 \frac{\delta z}{z} = \frac{\delta x}{x} + \frac{\delta y}{y}$$

giving $\delta z / z$ as required.

Quotients

Quotients can be treated as products in which some of the powers are negative. As before, the maximum value of δz is obtained by neglecting negative signs in the differential and combining all the terms additively.

If a function other than those already listed is encountered, some kind of differentiation usually works. It is frequently convenient to differentiate an equation implicitly, thereby avoiding the requirement to calculate the unknown quantity explicitly as a function of the other variables. For example, consider the thin-lens equation. If we had made measurements of the object distance s and the image distance s' for a thin lens with the intention of calculating a value for its focal length, f, the equation we would use is

$$\frac{1}{f} = \frac{1}{s} + \frac{1}{s'}$$

To obtain the uncertainty in f, we can differentiate the equation implicitly to obtain

$$-\frac{df}{f^2} = -\frac{ds}{s^2} - \frac{ds'}{s'^2}$$

It is now possible to calculate df/f directly and more easily than by writing f explicitly as a function of s and s', and differentiating. In this way we can prepare a formula for the uncertainty into which all the unknowns can be inserted directly. Make sure that appropriate signs are used so that all contributions to the uncertainty add positively to give outer limits of possibility for the answer.

If the function is so big and complicated that we cannot obtain a value for δz in general, we can always take the measured values, x_0, y_0, and so on, and work out z_0. We can then work out two different answers, one using the actual numerical values of $x_0 + \delta x$, $y_0 + \delta y$ (or $y_0 - \delta y$ if appropriate), and so forth to give one of the outer values of z, and the other using $x_0 - \delta x$, and so on. These two values correspond to the outer limits on z, and we know the value of δz.

2–10 COMPENSATING ERRORS

A special situation can appear when compound variables are involved. Consider, for example, the well-known relation for the angle of minimum deviation D_m for a prism of refractive index n and vertical angle A:

$$n = \frac{\sin \frac{1}{2}(A + D_m)}{\sin \frac{1}{2} A}$$

If A and D_m are measured variables with uncertainties δA and δD_m, the quantity n will be the required answer, with an uncertainty δ_m. It would be fallacious, however, to calculate the uncertainty in $A + D_m$, then in $\sin \frac{1}{2}(A + D_m)$, and combine that with the uncertainty in $\sin \frac{1}{2} A$, as if the function were a quotient of two independent variables. This can be seen by thinking of the effect on n of an increase in A. *Both* $\sin \frac{1}{2}(A + D_m)$ and $\sin \frac{1}{2} A$ increase, and the change in n is not correspondingly large. The error lies in applying the methods of the preceding sections to variables, such as A and $A + D_m$, that are not independent. The cure is either to reduce the equation to a form in which the variables are all independent or else to go back to first principles and use Equation (2–2) directly. Cases that involve compensating errors should be watched carefully, because if they are treated incorrectly, they give rise to errors in uncertainty calculations that are hard to detect.

2–11 SIGNIFICANT FIGURES

Because computations tend to produce answers consisting of long strings of numbers, we must be careful to quote the final answer sensibly. If, for example, we are given the voltage across a resistor as 15.4 ± 0.1 volts and the current as 1.7 ± 0.1 amps, we can calculate a value for the resistance. The ratio V/I comes out on my calculator as 9.0588235 ohms. Is this the answer? Clearly not. A brief calculation shows that the absolute uncertainty in the resistance is close to 0.59 ohms. So, if the first two places of decimals in the value for the resistance are uncertain, the rest are clearly meaningless. A statement like $R = 9.0588235 \pm 0.59$ ohms is, therefore, nonsense. We should quote our results in such a way that the answer and its uncertainty are consistent, perhaps something like $R = 9.06 \pm 0.59$ ohms. But is even this statement really valid? Remember that the originally quoted uncertainties for V and I had the value ± 0.1, containing one significant figure. If we do not know these uncertainties any more precisely than that, we have no right to claim two significant figures for the uncertainty in R. Our final, valid, and self-consistent statement is, therefore, $R = 9.1 \pm 0.6$ ohms. Only if we had a good reason to believe that our original uncertainty was accurate to two significant figures, could we lay claim to two significant figures in the final uncertainty and a correspondingly more precisely quoted value for R. In general terms, we must make sure that our quoted values for uncertainty are consistent with the precision of the basic uncertainties, and that the number of quoted figures in the final answer is consistent with the uncertainty of that final answer. We must avoid statements like $z = 1.234567 \pm 0.1$ or $z = 1.2 \pm 0.000001$.

PROBLEMS

1. I use my meter stick to measure the length of my desk. I am sure that the length is not less than 142.3 cm and not more than 142.6 cm. State this measurement as a central value ± uncertainty. What is the relative uncertainty of the measurement?

2. I read a needle-and-scale voltmeter and ammeter and assess the range of uncertainty visually. I am sure the ammeter reading lies between 1.24 and 1.25 A and the voltmeter reading between 3.2 and 3.4 V. Express each reading as a central value ± uncertainty and evaluate the relative uncertainty of each measurement.

3. My digital watch gives a time reading as 09:46. What is the absolute uncertainty of the measurement?

4. If I can read a meter stick with absolute uncertainty ±1mm, what is the shortest distance that I can measure if the relative uncertainty is not to exceed (a) 1%, (b) 5%?

5. I use a thermometer graduated in fifths of a degree Celsius to measure outside air temperature. Measured to the nearest fifth degree, yesterday's temperature was 22.4° Celsius and today's is 24.8° Celsius. What is the relative uncertainty in the temperature difference between yesterday and today?

6. The clock in the lab has a seconds hand that moves in one-second steps. I use it to measure a certain time interval. At the beginning of the interval it reads 09:15:22 (hours:minutes:seconds), and at the end it reads 09:18:16. What is the relative uncertainty of the measured time interval?

7. For the desk mentioned in Problem 1, I measure the width, and I am sure the measurement lies between 78.2 cm and 78.4 cm. What is the absolute uncertainty of the calculated area of the desktop?

8. In measuring the resistance of a resistor, the voltmeter reading was 15.2 ± 0.2 V and the ammeter reading was 2.6 ± 0.1 A. What is the absolute uncertainty of the resistance calculated using the equation $R = V / I$?

9. A simple pendulum is used to measure the acceleration of gravity using

$$T = 2\pi\sqrt{\frac{\ell}{g}}$$

The period T was measured to be 1.24 ± 0.02 s and the length to be 0.381 ± 0.002 m. What is the resulting value for g with its absolute and relative uncertainty?

10. An experiment to measure the density, d, of a cylindrical object uses the equation

$$d = \frac{m}{\pi r^2 \ell}$$

where

$$m = \text{mass} = 0.029 \pm 0.005 \text{ kg}$$
$$r = \text{radius} = 8.2 \pm 0.1 \text{ mm}$$
$$\ell = \text{length} = 15.4 \pm 0.1 \text{ mm}$$

What is the absolute uncertainty of the calculated value of the density?

11. The focal length, f, of a thin lens is to be measured using the equation

$$\frac{1}{f} = \frac{1}{s} + \frac{1}{s'}$$

where

$$s = \text{object distance} = 0.154 \pm 0.002 \text{ m}$$
$$s' = \text{image distance} = 0.382 \pm 0.002 \text{ m}$$

What is the calculated value for focal length, its absolute uncertainty, and its relative uncertainty?

12. A diffraction grating is used to measure the wavelength of light using the equation

$$d \sin\theta = \lambda$$

The value of θ is measured to be $13° \ 34' \pm 2'$. Assuming that the value of d is 1420×10^{-9} m and that its uncertainty can be ignored, what are the absolute and relative uncertainties in the value of λ?

13. A value is quoted as 14.253 ± 0.1. Rewrite it with the appropriate number of significant figures. If the value is quoted as 14.253 ± 0.15, how should it be written?

14. A value is quoted as $6.74914 \pm 0.5\%$. State it as a value $\pm$ absolute uncertainty, both with the appropriate number of significant figures.

3

Statistics of Observation

3–1 STATISTICAL UNCERTAINTY

In the preceding chapter we considered measurements in which the uncertainty could be estimated by personal judgment. In these, supposing that we have judged the situation accurately, repeated measurements should give consistent answers. Sometimes, however, systems behave in a different way and repeated measuring gives clearly *different* answers. For example, if we are using a particle detection and counting system to measure the activity of a radioactive source and we decide, with given geometry, to obtain the number of counts in a 10-second interval, we find that the results obtained by counting in successive 10-second intervals are not the same. We can encounter the same situation in measurements that involve visual judgment. If, for example, we wish to find the image formed by a thin lens, we may be unable to judge the position of the image accurately enough to obtain repeatedly the same reading on a good, finely divided distance scale. In many other systems, too, measurements show random fluctuation. Whether the fluctuation is intrinsic to the system under investigation (as in the radioactive source, where it arises from the basic nature of radioactive decay) or whether the variation arises from difficulty we have in making the measurement, we must find out how to make sensible statements about measured values that show such fluctuation.

What kind of statement will it be possible to make? No longer is it possible for us to make such statements as we made earlier. They had the form "I am virtually certain that the answer lies within the interval ...."

Now, when measured values appear seemingly at random along a scale, we are unable to identify the edges of an interval within which we can be almost certain our answer lies. In fact, apart altogether from the impossibility of obtaining "right" answers, we find that the difficulty lies not so much in constructing sensible answers as in knowing the sensible questions to ask. We discover that the only sensible questions involve, as before, intervals along the scale of values. This time, however, we interpret these intervals in terms of *probability* rather than certainty. Our search for a solution is fairly lengthy, but at the end the answer turns out to be simple and elegant.

To start the search, we go back to the basic situation. Assume that we have made a single measurement and that we have made the measurement a second time to check our work. This time we obtain a different value. What are we supposed to do? We have no way of saying that one value is "right" and the other "wrong." Which one should we choose to be "right"? In response to this ambiguity the natural reaction iws to try a third time, hoping, perhaps, that the third reading will confirm one or other of the first two. Very likely it will not be so obliging and will simply add to the confusion by supplying a third possibility. Faced with growing complexity, we can decide to keep on making measurements to see what happens. Suppose that our curiosity has prompted us to make a substantial number of repeated measurements, say 100, and we now ask: What is the answer? As was mentioned earlier, it is more significant to ask: What is the question? That depends very much on the use to which we wish to put the measurements. If we are measuring the position of an optical image, we may need a value that we can use in the design of some piece of optical equipment. If we are measuring the activity of a radioactive source, we may wish to make a guess at the number of counts that will be observed in a certain 10-second interval tomorrow. A sociologist counting political opinions wishes to predict the outcome of the next election, and so on. There is no single question and no unique answer. The treatment we give our fluctuating numbers depends on circumstances. We now consider some of the possibilities.

3–2 HISTOGRAMS AND DISTRIBUTIONS

Assume that we have made 100 measurements of some quantity and that we must now report our results. The first response to the question—What did you obtain?—is the rather feeble reply: "I made the measurement 100 times and here are the 100 answers." This statement may be clear and free from the possibility of misinterpretation, but it is hardly helpful. Our audience will find it difficult to make any sense out of a plain list of numbers, and questions will naturally arise, such as: Is there anything systematic about the numbers? Are there any regularities? Do any appear more frequently than others? And so on. To show the characteristics of the measurements more clearly, some kind of graphic display would clearly be helpful.

One common mode of presentation is the **histogram**. To construct this diagram, divide the scale along which the measurements are spread into intervals, and count the readings that fall within each interval. Then plot these numbers on a vertical scale against the intervals themselves. It is conventional to use a bar diagram to indicate the number of readings; and the result will be similar to Figure 3–1. At once we improve our comprehension of the measurements enormously, because we can see at a glance how the values are distributed along the scale. This distribution is the key to satisfactory interpretation of the measurements. Usually we find that the readings tend to occur more frequently in the middle of the range If this is so and we are unable to make any other sensible statement, we can always content ourselves with the simple assertion that the observations have central tendency. This may suffice, and when we have drawn the histogram we may be able to stop.

Many spreadsheet programs for computers such as Lotus 1-2-3, QuattroPro, and thge like, have built-in programs for calculating and displaying frequency distributions and histograms. These allow convenient and rapid statistical analysis of sets of observations, and it is good to use any available opportunity to practice with them. Their very convenience, however, can be deceptive, and it is very important to know by personal experience what kind of calculation the machine is doing. For this purpose start by first doing a few calculations by hand to see the way the numbers work. The examples at the end of this chapter can be used for this purpose.

Many results from measuring processes are presented simply by offering the histogram of the observations; readers can view the distribution and draw their own conclusions.

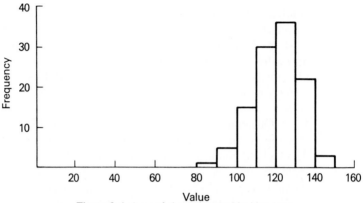

Figure 3–1 A set of obervations and its histogram.

85	109	114	121	127	131
92	109	114	121	127	132
96	110	114	122	127	133
97	110	115	122	127	134
97	111	116	122	128	134
97	111	116	122	128	134
100	111	116	122	128	134
101	111	117	123	128	135
101	111	117	123	128	136
102	112	118	123	128	137
102	112	118	123	130	137
103	112	119	123	130	137
103	113	119	124	130	144
105	113	120	124	130	148
106	113	120	124	130	149
106	113	120	125	130	
107	113	120	125	131	
108	113	121	125	131	
108	114	121	126	131	

3–3 CENTRAL VALUES OF DISTRIBUTIONS

Frequently, however, we wish to go further. If, for example, we could find some single number that could be used as a substitute for the whole distribution, it might simplify the reporting of our results. As candidates for a single number to represent the distribution as a whole, there are several possibilities, and we choose one or another, depending on the future use of the information. The various possibilities are mode, median, and mean.

Mode

Many distributions have a peak near the center. If the peak is well defined, the value on the horizontal scale at which it occurs is called the **mode** of the distribution. Whenever we wish to draw attention to such central concentration in our measured values, we quote the modal value. Sometimes a distribution shows two peaks; we call it a **bimodal** distribution and quote the two modal values.

Median

If we place all our readings in numerical order and divide the set into two equal parts, each containing the same number of readings, the value at which the dividing line comes is called the **median**. Because it is obvious that areas under distribution graphs represent numbers of observations (the left-hand bar in Figure 3–1 represents 5 observations, the second from the left represents 9, so that the two together represent 14, and so on), the median is that value at which a vertical line divides the distribution into two parts of equal area. The median is frequently quoted in sociological work; people talk about median salaries for certain groups of employees, for example.

Mean

The third of the commonly quoted numbers is the familiar arithmetic average, or **mean**. For a group of N observations, x_i, the mean $\bar{x}$ is defined by

$$\bar{x} = \frac{\Sigma x_i}{N} \qquad (3-1)$$

We shall discover that for our purposes the mean is the most useful of the three quantities we have defined.

Notice that for a symmetrical distribution the mean, median, and mode all coincide at the center of the distribution. On the other hand, if the distribution is not symmetrical, each has a separate value. For the histogram shown in Figure 3–1, the values of the mean, median, and mode are shown in Figure 3–2, which illustrates their relationship to the distribution as a whole.

If the distribution is markedly asymmetric, the difference between the mode, median, and mean can be substantial. Consider, for example, the distribution of family income in a country. The presence of the very wealthy, although they are relatively few in number, has an effect on the mean that counterbalances many members of the population at the low end of the salary scale. The mode

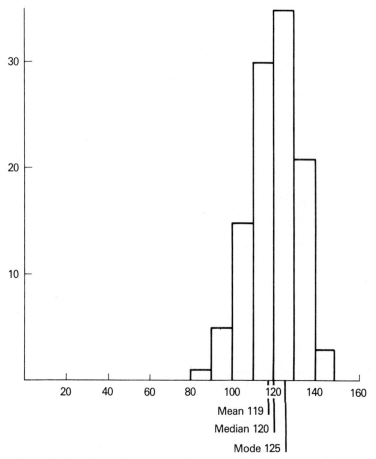

Figure 3-2 The relationship between a histogram and its mean, median, and mode.

and the mean thus differ substantially. This example illustrates the care required in interpreting quoted statistics; people who quote statistics frequently do so in the way that best suits their particular purpose.

3–4 THE BREADTH OF DISTRIBUTIONS

Let us now turn to the question: To what extent is our chosen number representative of the distribution as a whole? That is, how reliable is it to use a single number as a substitute for a whole distribution? In answering that question, we have at the present stage no justification to offer for the procedures that will be described. We rely, instead on an intuitive feeling that narrow distributions give us more confidence in the results than do broad distributions.

Let us, therefore, construct a quantity that is a measure of the breadth of the distribution. We could invent many such quantities, but, for reasons that need not concern us at the moment, we define a quantity that is almost universally used. Consider a set of N measurements x_i, of which the mean is $\bar{x}$. We define the **standard deviation** of the set of values, S, to be

$$S = \sqrt{\frac{\Sigma(\bar{x} - x_i)^2}{N}} \tag{3--2}$$

The definition is to some extent arbitrary, for in defining a measure of the breadth of the distribution we could have chosen other powers to which the quantity $(\bar{x} - x_i)$ could be raised, and we could have chosen other denominators. There are, however, reasons for these choices; these reasons and the significance of the standard deviation will become clear shortly.

To calculate the standard deviation for a set of numbers, it is important to interpret the quantities in Equation 3–2 correctly. Each of the quantities $\bar{x} - x_i$ is the difference between the mean and an individual number in the set. That difference may be positive or negative. When we square these differences, we obtain a series of positive numbers whose sum is dependent on the breadth of the distribution and provides the numerator in the expression for S. Almost all spreadsheet programs for computers and many calculators have built-in facilities for calculating standard deviations, and these greatly reduce the effort required. As was mentioned earlier with regard to distribution curves, however, it is important to become personally familiar with the way the numbers work. The examples at the end of the chapter, therefore, should be worked out by hand so that we may later know what calculators or computers are doing.

We can pause at this stage to summarize the progress so far. If we have made repeated measurements of a quantity and wish to state the result in numerical terms, we can do a number of things: (1) we can show the histogram, (2) we can quote the mode, median, or mean as a measure of the location of the distribution, and (3) we can quote the standard deviation as a measure of the breadth of the distribution. We sometimes leave the outcome of a measuring process in this form; the quantities involved are universally understood, and the procedure is acceptable.

For our present purpose, however, we seek more detailed, numerical interpretation of the quoted values.

3–5 SIGNIFICANCE OF THE MEAN AND STANDARD DEVIATION

In this and the following sections, for reasons that will become clear, we ignore the mode and median and restrict ourselves to numerical interpretation of the mean and the standard deviation. Because the presence of random fluctuation has denied us the opportunity to identify a realistic interval within which we can feel certain our answer lies, we must alter our expectations of the measuring process. As said before, it is not so much a matter of obtaining sensible answers to questions as of knowing the sensible questions to ask. Specifically, it is not sensible to ask: What is the right answer? It is not even sensible to ask: Having made 100 observations of a quantity, what shall I obtain when I make the measurement the next time? The only sensible questions involve not certainty but probability, and several different questions about probabilities are possible.

For example, we could ask: What is the probability that the 101st reading will fall within a certain range on our scale of values? That is a sensible question, and sensible answers can easily be imagined. If, for example, of our 100 original readings, a certain fraction of the values fell within some particular range, we might feel justified in choosing that fraction as the probability that the 101st observation will fall within that interval. This would not be an unrealistic guess, and we could attempt a standardized description of our distribution by quoting the fraction of the total number of readings that fall within various specified intervals. This would satisfactorily convey information about our set of readings to other people, but a major problem appears when we discover that our answers for these probabilities are specific to our particular histogram. If we were to make another series of 100 readings, holding all the conditions the same as they were before in the hope of obtaining the same histogram, we would be disappointed. The new histogram would not duplicate the first exactly. It might have similar general characteristics with respect to location and breadth, but its detailed structure would not be the same as before, and we would obtain different answers to questions about probabilities.

How, then, are we going to find answers to our questions that have some kind of widely understood numerical significance? One solution is to abandon the attempt to describe our particular histogram and to start talking about defined theoretical distributions. These may have the disadvantage of uncertain relevance to our particular set of observations, but there is the enormous advantage that, because they are defined theoretical constructs, they have properties that are definite, constant, and widely known. Many such theoretical distributions have been constructed for special purposes, but we restrict ourselves to one only, the **Gaussian**, or **normal**, distribution.

We use the Gaussian distribution to interpret many kinds of physical measurement, partly because the mechanical circumstances of many physical measurements are in close correspondence with the theoretical foundations of the Gaussian distribution, and partly because experience has shown that Gaussian statistics do provide a reasonably accurate description of many real events. For only one common type of physical measurement is another distribution more appropriate. In counting events like radioactive decay we must use a distribution called the Poisson distribution, but even for it the difference from Gaussian statistics becomes significant only at low counting rates. Further information about Poisson statistics can be found in books describing experimental methods in nuclear or high-energy physics. Apart from these special cases, we can feel relatively confident that Gaussian statistics can be usefully applied to most real measurements. Always remember, however, that unless we actually test our measurements for correspondence with the Gaussian distribution, we are making an *assumption* that Gaussian statistics are applicable, and we should remain alert to any evidence that the assumption may be invalid.

3–6 GAUSSIAN DISTRIBUTIONS AND SAMPLING

Even if to use it successfully we need not know very much about the origins of the Gaussian distribution, it is interesting to understand why its axiomatic foundations make it particularly relevant to many physical measurements. The equation for the Gaussian distribution can be derived from the assumption that the total deviation of a measured quantity, x, from the unperturbed value X that would be obtained in the absence of perturbing fluctuations is the consequence of a large number of small fluctuations that occur randomly in positive and negative directions. To construct a simple model of such a situation, suppose that there are m such contributions to the total deviation, each of equal magnitude a and equally likely to be positive or negative. Any individual measured value, x, therefore differs from X by an amount that contains a random number of positive contributions to the deviation, and a corresponding number of negative contributions. If we repeat the measuring process many times, therefore, we obtain a set of values that range from $X + ma$ for a measurement in which all the fluctuations happened to be positive simultaneously, to $X - ma$ if the same happened in the negative direction. Either of these possibilities is most unlikely. If perturbations occur randomly, it is much more likely that they will occur in a mixture of positive and negative contributions, and so the total measured quantity x is more likely to be found near the middle of the range of possibilities than at the ends.

Such a situation, in which we have a random summation of positive and negative quantities, is similar to the so-called random walk. In this, we depart from a starting point by deciding to take a step backward or forward, depending on whether a coin that we toss comes up heads or tails. It can easily be demonstrated that our most probable location after any number of steps is the position from which we started. It is the same with the measurements that have been perturbed by random perturbations. The most probable sum of the perturbations is zero, meaning that for repeated measurements, x, the most common values are in the vicinity of X. The distribution curve, therefore, has a peak in the middle, is symmetrical, and declines smoothly to zero at $X + ma$ and $X - ma$. If this concept is taken to the limiting case in which an infinite number of infinitesimal deviations contribute to the total deviation, the curve has the form shown in Figure 3–3. Treating the curve solely from the mathematical point of view for the moment, we can easily prove that its equation can be written

$$y = Ce^{-h^2(x-X)^2} \tag{3-3}$$

We need not be concerned about the derivation of this equation; for our purposes, it is sufficient to know the principles on which it is based and the resulting properties of the function. If we wish to know more about the origins of the function, the full derivation is in Appendix 1.

In the equation the constant C is a measure of the height of the curve, since $y = C$ for $x = X$ at the center of the distribution. The curve is symmetrical about $x = X$ and approaches zero asymptotically. The quantity h obviously governs the width of the curve, because it is only a multiplier on the x scale. If h is large, the curve is narrow and high in relation to its width; if small, the curve is low and broad. The quantity h clearly must be connected with the standard deviation of the distribution. Now that we are dealing

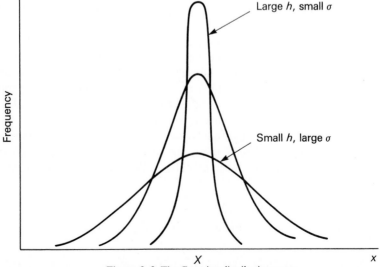

Figure 3–3 The Gaussian distribution curve.

with a mathematically defined theoretical distribution instead of a finite set of real observations, we modify our terminology a little. We used the latin letter S to denote the standard deviation of a finite set of *real* observations. We use the greek letter σ to represent the standard deviation that can be calculated for a *mathematically defined* distribution, such as the Gaussian function. For the Gaussian distribution, the relationship between its standard deviationσ and the geometrical measure of the width h is

$$\sigma = \frac{1}{\sqrt{2}h} \qquad (3\text{--}4)$$

Now that we have a definite equation for the distribution, all the original ambiguity about interpreting the standard deviation in terms of probability disappears. We now have definite, unique, and permanent values. For example, the area enclosed within the interval $X \pm \sigma$ for a Gaussian distribution is 68% of the total area under the curve, and within the interval $X \pm 2\sigma$ it is 95%, and this is so for *all* Gaussian distributions. The relation between the σ values and areas on the distribution curve is shown in Figure 3–4 by the lines drawn vertically at intervals of 1σ and 2σ from the central value.

It is very comforting to have such definite numbers, because we can say definitely that any particular value in a Gaussian set has a 68% chance of falling within the interval $X \pm \sigma$ and a 95% chance of falling within $X \pm 2\sigma$,

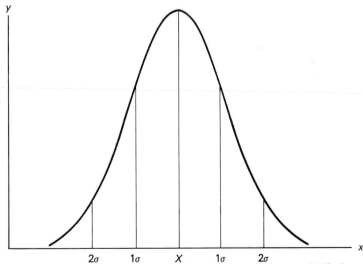

Figure 3–4 The relationship of 1σ and 2σ limits to the Gaussian distribution.

and we shall have the occasion to use these probability values repeatedly. A more extensive account of the mathematical properties of the Gaussian distribution is in Appendix 1.

3–7 RELATION BETWEEN GAUSSIAN DISTRIBUTIONS AND REAL OBSERVATIONS

The results given in the preceding section provide useful, precise methods for interpreting means and standard deviations, but a problem arises when we apply such thoughts to real measurements. Numbers like 68% and 95% refer specifically to a theoretical construct, the Gaussian distribution. When we engage in a real measuring process, all we have is one, or at most a few, actual measurements of our desired quantity. We have at first no way of knowing which Gaussian distribution, with attached values of X and σ, is appropriate to our observations. So what are we to do? The answer lies in a concept that provides a bridge between the world of theoretical constructs and the world of real measurements. For the particular circumstances of our measurement, we invent the concept of the infinite set of measurements that could be made. For rather obvious reasons, this infinite set of measurements will never be made, but the concept enables us to interpret our real measurements. The construct is called the **universe**, or **population**, for that particular measurement. Once we have made, say, 100 measurements with a particular apparatus, we tend to feel that nothing exists but our 100 values. We must now invert our thinking and view the measurements as a **sample** of the infinitely large universe, or population, of measurements that could be made. The universe, however, is permanently inaccessible to us; we shall never know the universe distribution or its mean or its standard deviation. Our task is to construct *inferences* about these quantities from the definitely known properties of our sample.

We do this on the basis of some assumptions. First, we assume that the universe distribution is Gaussian, and we call the universe mean X and the universe standard deviation σ. This assumption enables us to make statements such as: If we make just one measurement with our equipment, that one measurement has a 68% chance of falling within the interval $X \pm \sigma$ and a 95% chance of falling within $X \pm 2\sigma$. This seems like an encouragingly exact and explicit statement, but it has an overwhelming defect; we do not (and never shall) know the values of X and σ. In other words, having made only one observation of a quantity that is subject to random fluctuation, we have gained practically nothing. We can say only that our single measured value has a 68% chance of falling within something of somewhere, which is not too helpful. Our only hope lies in obtaining some information, even if uncertain, about the universe distribution. As already mentioned, we are never going to be able to determine the universe distribution exactly, because that would require an infinite number of readings. We can only hope that, if we repeat our measuring process to obtain a sample from the universe, that sample will enable us to make some estimate of the universe parameters.

Because we make the basic assumption that the universe distribution is a mathematical, defined function (whether Gaussian or some other equally well-defined distribution), we can evaluate mathematically the properties of samples taken from that universe and compare the sample properties with those of the distribution of single observations. We simply state these properties of samples without proof. The reader who is curious about the mathematical derivation of these results is encouraged to turn to the standard texts on statistics, in which there are sections dealing with sampling theory.

The properties of samples become clear if we consider the concept of repeated sampling. Consider that with a certain piece of apparatus we make 100 observations. This is our first sample; let us calculate its mean and standard deviation and record them. Now let us make another set of 100 observations and record for it the mean and standard deviation. We continue such repetition until we have an infinite number of samples, each with its own mean and standard deviation, and we then plot the distribution curves of these sample means and of the sample standard deviations. Of course, we shall never carry out a process like this with actual observations, but knowing the mathematical function for our original universe of single readings, we can simulate such repeated sampling mathematically and so derive the properties of the samples in comparison with those of the original universe of single readings. The results of such calculations of the distribution of sample means and sample standard deviations are shown in Figure 3–5 and Figure 3–6, and they are described in the following sections.

3–8   SAMPLE MEANS AND STANDARD DEVIATION OF THE MEAN

If the universe distribution of single readings is Gaussian, the theory of sampling shows that the distribution of sample means is also Gaussian. The distribution of sample means has two other very important properties. First, it is centered on X, the center of the original distribution of single readings; second, it is narrower than the original distribution. This narrowness is highly significant because it demonstrates immediately the improvement in precision that comes from samples as opposed to single readings; the means of samples cluster more closely around the universe mean than do single readings. The reduced scatter of sample means is represented by an important

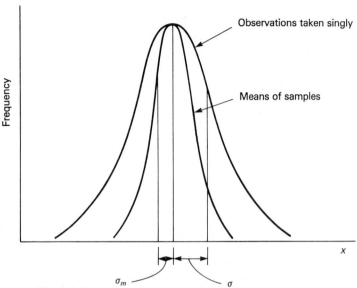

Figure 3–5 Distribution curve of single observations and sample means. (Note that the vertical scale for the two curves is not the same. They have been plotted with a common peak value solely for purposes of illustration.)

quantity—the standard deviation of the distribution of sample means. This quantity is called the **standard deviation of the mean**; its symbol is σ_m and sampling theory gives its value as

$$\sigma_m = \frac{\sigma}{\sqrt{N}} \tag{3–5}$$

where N is the number of readings in the sample. This result gives us the opportunity to make a very important statement. A particular sample mean has a 68% chance of falling within the interval $X \pm \sigma_m$, and a 95% chance for the interval $X \pm 2\sigma_m$. These intervals are smaller than the corresponding intervals for single readings, and they supply a numerical measure of the improved precision that is available from sampling.

Notice that the statement about sample means, although precise and important, still does not help us much, because it still involves the unknown quantities X and σ The resolution of this difficulty and the significance of the standard deviation of the mean will become clear soon. In the meantime we turn our attention briefly to the other important property of samples—the distribution of sample standard deviations.

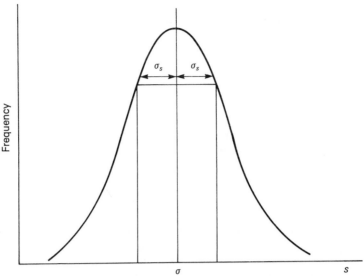

Figure 3–6 The distribution of sample standard deviations.

3–9 SAMPLE STANDARD DEVIATION

It can be proved by sampling theory that the sample standard deviations also fall on a Gaussian distribution that is centered on the value of the universe standard deviation σ. The distribution is illustrated in Figure 3–6. As will become clear, however, the variance of the sample standard deviations will not concern us as much as the variance of sample means, and we postpone to Section 3–11 further discussion of the variance of sample standard deviations.

3–10 APPLICATION OF SAMPLING THEORY TO REAL MEASUREMENTS

The sample properties just presented are interesting, but how do they help us when we do not have access to the actual distributions, either for sample means or sample standard deviations? When making real measurements, we obtain our lone sample with its mean and standard deviation, and we have no idea how these values relate to the universe values. The problem, therefore, is to find a connection between the theoretical results and the sample properties that allows us to infer the universe properties from the sample values. We cannot expect to obtain exact information. In addition, we must make one basic, obviously imprecise assumption. Assume that our single number, the standard deviation of our sample, provides us with a value for the universe standard deviation. In fact, it can be proved that the "best estimate" (a mathematically defined term) of the universe standard deviation is given by the quantity

$$S = \sqrt{\frac{\Sigma(\bar{x} - x_i)^2}{N - 1}} \qquad (3\text{--}6)$$

This quantity is only slightly different from the original value for the standard deviation of a set of observations. The N in the denominator of the original expression has been replaced by $N - 1$, and the difference between the two quantities is significant for only small values of N. In the future, when we talk about a sample standard deviation, we shall assume that we are using the equation in the new form and that we are really talking about the "best estimate" of the universe value σ.

Accepting the sample standard deviation as the best estimate of σ, we are now able to make a definite statement about the single sample. We can rephrase Equation (3–5) and define

$$S_m = \frac{S}{\sqrt{N}} \qquad (3\text{--}7)$$

as our standard deviation of the mean, now a known quantity obtained from our real sample. We can now say: The sample mean $\bar{x}$ has a 68% chance of falling within the interval $X \pm S_m$ and a 95% chance of falling within $X \pm 2S_m$. This statement is close to what we want, but it is not yet completely satisfactory. It tells us something about a quantity that we know, $\bar{x}$, in terms of a quantity that we do not know, X. We really want the statement to be the other way around; we want to be able to make an assertion about the unknown, X, in terms of a quantity, $\bar{x}$, of which we do know the value. Fortunately, it is possible to prove that the above statement about probabilities can be inverted to yield the desired result. We obtain thereby the statement toward which we have been working ever since we started the discussion of the statistics of fluctuating quantities. The final statement is: There is a 68% chance that the universe mean, X, falls within the interval $\bar{x} \pm S_m$. and a 95% chance that it falls within the interval $\bar{x} \pm 2S_m$. This is now, finally, a statement about the unknown quantity, X, in terms of wholly known quantities, $\bar{x}$ and S_m. Along the scale of x values, we now have a real and known interval between $\bar{x} - S_m$. and $\bar{x} + S_m$, and we know that there is a 68% chance that the desired quantity X lies within this interval.

This statement provides us with the answer we have been seeking and brings us as close as we can come to exact information about the unperturbed value of the measured quantity. It is worth becoming familiar with the arguments that have been given in the preceding sections; there is more to measurement than simply making a few measurements and "taking the average" just because it seems to be the right thing to do. We should understand fully the significance of what we are doing.

3–11   EFFECT OF SAMPLE SIZE

Clearly, in any sampling process, the larger the sample, the more precise the final statements. Even though the precision of a mean value increases only as the square root of the number of observations in the sample [Equation (3–5)], it does increase, and larger samples have more precise means. There may, however, be limitations of time or opportunity, and we cannot always obtain samples of the size we would like. Usually, a compromise must be sought between the conflicting demands of precision and time, and good experiment design incorporates this compromise into the preliminary planning. Nevertheless, it may occasionally be necessary to be content with small samples. In this undesirable eventuality, we should be aware of the magnitude of the resulting loss of precision. There is, first, the influence on the value of the standard deviation of the mean; the smaller N is, the larger the value of S_m, and the longer the interval on the x scale that has the 68% chance of containing the universe value X.

Second, for small samples, we must place declining faith in the use of the sample standard deviation S as the best estimate of the universe value σ. To illustrate this, recall the distribution curve for sample standard deviations shown in Figure 3–6. It is worth asking: Given the existence of this distribution, how good is our best estimate of the universe standard deviation, and how does it vary with sample size? The answer must be based on the width of the distribution of sample standard deviations, and so we should calculate the standard deviation of this distribution. It is called the **standard deviation of the standard deviation**. (This process could obviously go on indefinitely, but we stop at this stage.) The value of the standard deviation of the standard deviation, calculated mathematically by sampling theory on the basis of the equation of the Gaussian distribution, is

$$\sigma_S = \frac{\sigma}{\sqrt{2(N-1)}} \qquad (3\text{–}8)$$

The breadth of the distribution of sample standard deviations is thus related to its central value σ by the numerical factor $1/\sqrt{2(N-1)}$. As one might expect, therefore, the accuracy of the sample standard deviation as the best estimate of the universe value depends on the sample size. For example, with a sample size of 10, Equation (3–8) shows that the S value from the sample has a 68% chance of falling within an interval of $\pm\sigma/\sqrt{18}$, approximately $\pm\sigma/4$, about the universe value σ. Correspondingly, the interval that has a 95% chance of containing the sample standard deviation is as wide as $\pm\sigma/2$ about the universe value σ. This does not represent high precision of measurement. We have, therefore, confirmation of the warning given earlier; statistical exercises with small samples should be undertaken only when no alternative exists. To provide an overall feeling for the reliability of σ estimates from samples of differing size, Table 3–1 contains some typical values of $\sqrt{2(N-1)}$ for various values of N.

These values are illustrated in Figure 3–7 for $N=3$, $N=10$, and $N=100$. The $\pm 1\sigma_S$ limits are marked on these curves, showing, for various sample sizes, the intervals within which there is a 68% probability that the single sample standard deviation lies. For values of N less than about 10, it is clear that the intervals for 68% or 95% probability become so large in comparison with the central value that it is almost pointless to attempt an estimate of σ. It is rarely worth attempting any kind of statistical analysis with samples containing fewer than about 10 observations. When reporting the outcome of statistical work, it is essential to quote the sample size. If we intend our values for the mean and standard deviation of the mean to be interpreted in accordance with the 68% and 95% prescription, we must give our readers the opportunity to judge the accuracy of our estimates.

TABLE 3–1 Accuracy of σ Estimates from Samples of Varying Size

68% Confidence		95% Confidence	
N	$\sqrt{2(N-1)}$	N	$\sqrt{2(N-1)}$
2	1.4	2	0.7
3	2.0	3	1.0
4	2.4	4	1.2
5	2.8	5	1.4
6	3.1	6	1.6
7	3.4	7	1.7
8	3.7	8	1.8
9	4.0	9	2.0
10	4.2	10	2.1
15	5.2	15	2.6
20	6.1	20	3.2
50	9.8	50	4.9
100	14.1	100	7.0

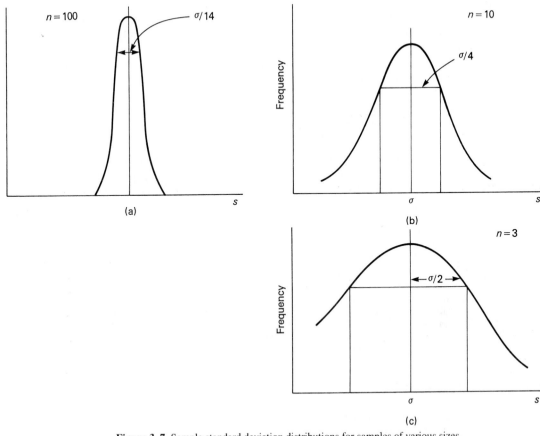

Figure 3–7 Sample standard deviation distributions for samples of various sizes.

3–12   STANDARD DEVIATION OF COMPUTED VALUES

In Chapter 2 we considered the uncertainty of computed values z, and we assumed that the uncertainty of the basic measurements constituted intervals within which we were almost certain that the values lay. We calculated the maximum range of variability of the computed answer on the pessimistic assumption that the errors in the various measured values had combined in a worst-case fashion to drive the computed answer as far away from the central value as it could go. We have already suggested that this represents an unrealistically pessimistic approach and that a more useful quantity would be a *probable* value for the uncertainty in z that is based on the various probabilities associated with deviation of the basic quantities x, y, and so on, from their central values. The limits given by this quantity will naturally be smaller than the $\pm\delta z$ that we calculated before, but we can hope to find actual numerical significance for them. Such statistical validity is available only if the uncertainties in x and y have statistical significance, and we assume in the following calculations that the measurements of x and y have been sufficiently numerous to justify a calculation of the standard deviations S_x and S_y. We hope now to calculate a value for S_z that will have the same significance for z values as S_x and S_y had for the values of x and y.

We must first ask what we mean by S_z. To construct an interpretation, we assume that the measuring process has given us N pairs of observations x, y that were obtained by repetition of the observing process under identical conditions (for example, the current through and the potential across a resistor that had been measured for the purpose of calculating the resistance R). Each pair of observations provides a value of z through some functional relation $z = f(x,y)$. Because repetition of the basic measurements yielded N pairs, we now have a set of N values of z. These are not identical because of fluctuations in the basic measurements x and y. The z values therefore fall on a distribution curve, and the quantity we require, S_z, is the standard deviation of this set of z values. These individual values of z may, of course, never be calculated separately because a simpler mode of calculating our final answer exists. We can calculate the means $\bar{x}$ and $\bar{y}$ of the sets of x and y values and obtain $\bar{z}$ directly by using the assumption (valid if S_x, S_y, and S_z are small compared, respectively, with $\bar{x}$, $\bar{y}$, and $\bar{z}$) that

$$\bar{z} = f(\bar{x},\bar{y})$$

Even if we never calculate them individually, that distribution of separate z values provides the significance of the S_z that we are about to calculate.

If we assume that the universes of the separate x, y, and z values have Gaussian distributions, the quantity σ_z (of which we are about to calculate the best estimate in terms of the various S values) has the usual significance (i.e., any particular z value has a 68% chance of falling within $\pm\sigma_z$ of the central value). As before, let

$$z = f(x,y)$$

and consider perturbations δx and δy that lead to a perturbation δz in the computed value of z. The value of δz will be given, as before, by

$$\delta z = \frac{\partial z}{\partial x}\delta x + \frac{\partial z}{\partial y}\delta y$$

This perturbation in z can be used to calculate a standard deviation for the N values of z. Each perturbation in z is equivalent to the difference between a mean and a measured value that appears in the definition of a standard deviation. Hence, the standard deviation of the set of z values is given by

$$S_z = \sqrt{\frac{\Sigma(\delta z)^2}{N}}$$

Thus

$$S_z^2 = \frac{1}{N}\Sigma\left(\frac{\partial z}{\partial x}\delta x + \frac{\partial z}{\partial y}\delta y\right)^2$$

$$= \frac{1}{N}\Sigma\left(\left(\frac{\partial z}{\partial x}\right)^2(\delta x)^2 + \left(\frac{\partial z}{\partial y}\right)^2(\delta y)^2 + 2\frac{\partial z}{\partial x}\frac{\partial z}{\partial y}\delta x\delta y\right)$$

$$= \left(\frac{\partial z}{\partial x}\right)^2\frac{1}{N}\Sigma(\delta x)^2 + \left(\frac{\partial z}{\partial y}\right)^2\frac{1}{N}\Sigma(\delta y)^2 + \frac{2}{N}\frac{\partial z}{\partial x}\frac{\partial z}{\partial y}\Sigma(\delta x\delta y)$$

But $\quad\dfrac{1}{N}\Sigma(\delta x)^2 = S_x^2 \qquad$ and $\qquad \dfrac{1}{N}\Sigma(\delta y)^2 = S_y^2$

Also, because δx and δy may be considered for the present purpose to be independent perturbations,

$$\Sigma(\delta x\delta y) = 0$$

Thus, finally,

$$S_z = \sqrt{\left(\frac{\partial z}{\partial x}\right)^2 S_x^2 + \left(\frac{\partial z}{\partial y}\right)^2 S_y^2} \tag{3-9}$$

If z is a function of more than two variables, the equation is extended by adding similar terms. Thus, if the components of a calculation have standard deviations with some degree of reliability, a value can be found for the probable uncertainty of the answer, where "probable" has real numerical significance.

The calculation has been performed in terms of the variance or standard deviation of the x and y distributions. In actual practice, however, we do not use the sample variance directly; we must calculate the best estimates of σ_x, σ_y, and so on, and in accordance with Equation (3–6), we use the modified value for standard deviation with denominator $N-1$ instead of N. The final result is then a best estimate for σ_z. The standard deviation of the mean for z is then calculated by direct use of Equation (3–5) and gives the limits that have a 68% chance of containing the unperturbed value of z.

Note that most experiments are not carried out in accordance with the restricted assumptions of the foregoing development. If, for example, we are studying the flow rate of water through a pipe, we would measure the flow rate, pipe radius, and pipe length independently and would choose the number of readings in each sample on the basis of the intrinsic precision of the measurement. We cannot, therefore, use Equation (3–9) directly, because the various S's are not compatible. The solution is to calculate the standard deviation of the mean for each of the elementary quantities first. If these are used in Equation (3–9), the result of the calculation is immediately a standard deviation of the mean for z.

3–13 STANDARD DEVIATION OF COMPUTED VALUES: SPECIAL CASES

Let us now apply Equation (3–8) to a few common examples. In all the following cases the various Ss are all assumed to be best estimates of the appropriate universe value σ.

Sum of Two Variables

If

$$z = x + y$$

then

$$\frac{\partial z}{\partial x} = 1, \qquad\qquad \frac{\partial z}{\partial y} = 1$$

and

$$S_z = \sqrt{S_x^2 + S_y^2}$$

Note that this result provides justification for Eq. (3–5). The mean value for the sample, $\sum (x_i) / N$, is just such a function as $z = x + y$, where x and y happen to be independent measurements of the *same* quantity. Thus, if

$$z = \frac{1}{N}(x_1 + x_2 + x_3 + \ldots)$$

$$\frac{\partial z}{\partial x_1} = 1/N, \qquad\qquad \frac{\partial z}{\partial x_2} = 1/N, \qquad\qquad \text{and so on,}$$

and

$$S_z = \sqrt{\left(\frac{1}{N}\right)^2 S_x^2 + \left(\frac{1}{N}\right)^2 S_x^2 + \dots}$$

$$= \sqrt{N S_x^2 / N^2}$$

$$= S_x / \sqrt{N}$$

which is the result we had earlier for the standard deviation of the mean.

Difference of Two Variables

If

$$z = x - y$$

$$\frac{\partial z}{\partial x} = 1, \qquad\qquad \frac{\partial z}{\partial y} = -1$$

but, again,

$$S_z = \sqrt{S_x^2 + S_y^2}$$

Recalling Section 2-8, we note that the earlier discussion of measured differences is still valid. The standard deviations in x and y combine additively, even though the quantity $x - y$ can have quite small values.

Product of Two Variables

If

$$z = xy$$

$$\frac{\partial z}{\partial x} = y, \qquad\qquad \frac{\partial z}{\partial y} = x$$

and

$$S_z = \sqrt{y^2 S_x^2 + x^2 S_y^2}$$

The specific value of S_z at any particular values of x and y, say x_0 and y_0, can be obtained by substituting x_0 and y_0 in this expression. As was the case for uncertainty in products, the equation is more clearly expressed in terms of relative values of S_z. We obtain

$$\frac{S_z}{z} = \sqrt{\frac{S_x^2}{x^2} + \frac{S_y^2}{y^2}}$$

Variables Raised to Powers

If

$$z = x^a$$

$$\frac{\partial z}{\partial x} = ax^{a-1}$$

and

$$S_z = \sqrt{a^2 x^{2(a-1)} S_x^2}$$
$$= ax^{(a-1)} S_x$$

Again, this result is more instructive and more easily remembered when expressed in terms of relative values:

$$\frac{S_z}{z} = \frac{ax^{(a-1)} S_x}{x^a}$$
$$= a \frac{S_x}{x}$$

The General Case of Powers and Products

If

$$z = x^a y^b$$

the results of the two preceding sections can obviously be extended to give the result

$$\frac{S_z}{z} = \sqrt{\left(\frac{aS_x}{x}\right)^2 + \left(\frac{bS_y}{y}\right)^2}$$

In contrast to the case of combined uncertainty, negative powers in the original function need not be given special consideration; in the equation for S_z powers occur in squared form and automatically make a positive contribution.

If a function other than those we have listed is encountered, the use of Equation (3–9) yields the desired result. Incidentally, we may note that, for a function of a single variable, Equation (3–9) reduces to the same form as for uncertainties, Equation (2-1). This correspondence could easily have been predicted for a situation in which we do not have the probability-based interplay between two or more variables.

3–14   COMBINATION OF DIFFERENT TYPES OF UNCERTAINTY

Unfortunately for the mathematical elegance of the development, we frequently require the uncertainty in a computed result to contain quantities having different types of uncertainty. We may require the uncertainty in a function

$$z = f(x, y)$$

in which, for example, x is a quantity to which have been assigned outer limits, $\pm \delta x$, within which we are "almost certain" that the actual value lies, and y is a quantity whose uncertainty is statistical in nature, a sample standard deviation, S_y, perhaps, or a standard deviation of the mean, $S_y / \sqrt{N}$. We require an uncertainty for z. The initial difficulty is even to define the uncertainty in z. We are trying to combine two quantities that in effect have completely different distibution curves. One is the standard Gaussian function; the other is a rectangle. This rectangle is bounded by the values $x_0 + \delta x$ and $x_0 - \delta x$ and is flat on top because the actual value of x is equally likely to be anywhere within the interval $x_0 \pm \delta x$. Any general method of solving this problem is likely to be far too complex for general use, but a simple approximation is obtainable by using the following procedure.

In the calculation for z we could use the sample mean, $\bar{y}$, for the y value, implying that the universe mean has an approximately two-thirds chance of falling within the interval, $\bar{y} \pm S_y / \sqrt{N}$. We, therefore, calculate limits for x that also have a two-thirds probability of enclosing the actual value. Because the probability distribution for x is rectangular, two-thirds of the area under the distribution curve is enclosed by limits that are separated by a distance equal to two-thirds of the total range of possibility (i.e., two-thirds of $2\delta x$. The total width of the region for two-thirds probability is therefore, $(4/3)\delta x$ and the uncertainty limits are $\pm(2/3)\delta x$.

The quantity $(2/3)\delta x$ is compatible with $S_y / \sqrt{N}$ because both refer to two-thirds probability. Equation (3–9) can now be used, inserting $(2/3)\delta x$ for the value of the standard deviation of the mean for x and $S_y / \sqrt{N}$ for the y function. This yields a value for uncertainty in z which can be interpreted in accordance with the two-thirds prescription. Note that the limits for 95% probability are not simply twice as wide as those for two-thirds probability; they must be calculated separately using the foregoing method.

3–15   REJECTION OF READINGS

One last practical property of distribution curves concerns outlying values. There is always the possibility of making an actual mistake, perhaps by misreading a scale or in accidentally moving an instrument between setting and reading. There is the temptation to assign some such cause to a single reading that is well separated from an otherwise compact group of values. This is a dangerous temptation because the Gaussian curve does permit values remote from the central part of the curve. Furthermore, once we admit the possibility of pruning the observations, it can become difficult to know where to stop. We are dependent, on the judgment of the experimenter. This is not unreasonable, because the experimenter knows more about the measurement than anyone else, but criteria for making the choices can be helpful. Many empirical "rules" for rejection of observations have been formulated, but they must be used with discretion. It would be foolish to use a rule to reject one reading that was just outside the limit set by the rule if there are other readings just inside it. There is also the possibility that extra information relating to the isolated reading was noted at the time it was made, and this can help us decide in favor of retention or rejection.

The guidance for making such decisions can be found in the properties of the Gaussian distribution. In a Gaussian distribution the probability of obtaining readings outside the 2σ limits is 5% (as we have seen before), outside 3σ limits it is approximately 0.3%, and outside 4σ limits the chance is no more than 6×10^{-5}. The decision to reject is still the responsibility of the experimenter, but we can say in general terms that readings that fall outside 3σ limits are likely to be mistakes and candidates for rejection. However, a problem can arise because of our lack of information about the universe of readings and its parameters X and σ. The better our knowledge of σ, the more confident we can be that any far-out and isolated reading arises from a genuinely extraneous cause such as personal error, malfunction of apparatus, and the like. Thus, if we make 50 observations that cluster within 1% of the central value and then obtain one reading that lies at a separation of 10%, we can be fairly safe in suggesting that this last reading did not belong to the same universe as the preceding 50. The basic requirement before any rejection is justified is confidence in the main distribution of readings. Clearly, there is no justification for taking two readings and then rejecting a third measurement on the basis of a 3σ criterion. Unless the case for rejection is completely convincing, the best course is to retain all readings, whether we like them or not.

It is wise also to remember that many of the greatest discoveries in physics had their origin in outlying measurements.

PROBLEMS

The following observations of angles (in minutes of arc) were made while measuring the thickness of a liquid helium film. Assume that the observations show random uncertainty, that they are a sample from a Gaussian universe, and use them in Problems 1 to 14.

34	35	45	40	46
38	47	36	38	34
33	36	43	43	37
38	32	38	40	33
38	40	48	39	32
36	40	40	36	34

1. Draw the histogram of the observations.

2. Identify the mode and the median.

3. Calculate the mean.

4. Calculate the best estimate of the universe standard deviation.

5. Calculate the standard deviation of the mean.

6. Calculate the standard deviation of the standard deviation.

7. (a) Within which limits does a single reading have a 68% chance of falling?

(b) Which limits give a 95% chance?

8. Within which limits does the mean have (a) a 68% chance, and (b) a 95% chance of falling?

9. Within which limits does the sample standard deviation stand (a) a 68% chance and (b) a 95% chance of falling?

10. Calculate a value for the constant h in the equation for the Gaussian distribution.

11. If a single reading of 55 had been obtained in the set, would you have decided in favor of accepting it or rejecting it?

12. Take two randomly chosen samples of five observations each from the main set of readings. Calculate their sample means and standard deviations to see how they compare with each other and with the more precise values obtained from the complete sample.

13. If the experiment requires that the standard deviation of the mean should not exceed 1% of the mean value, how many readings will be required?

14. If the standard deviation of the universe distribution must be known within 5%, how many readings will be required?

15. Repeated measurements of the diameter of a wire of circular cross section gave a mean of 0.62 mm with a sample standard deviation of 0.04 mm. What is the standard deviation for the calculated value of the cross-sectional area?

16. The wavelength of the two yellow lines in the sodium spectrum are measured to be 589.11×10^{-9} m and 589.68×10^{-9} m, each with a standard deviation of 0.15×10^{-9} m. What is the standard deviation for the calculated difference in wavelength between the two lines?

17. A simple pendulum is used to measure g using

$$T = 2\pi \sqrt{\frac{\ell}{g}}$$

Twenty measurements of T gave a mean of 1.82 s and a sample standard deviation of 0.06 s. Ten measurements of ℓ gave a mean of 0.823 m and a sample standard deviation of 0.014 m. What is the standard deviation of the mean for the calculated value of g?

4

Scientific Thinking
and Experimenting

4–1   OBSERVATIONS AND MODELS

In this chapter we briefly review the nature of scientific activity in the hope
that the procedures used in various types of experimenting will be seen to
arise naturally from the problems that are encountered. To understand the
nature of scientific thinking, it helps to go back to fundamentals and pretend
that we are inventing a new area of scientific study right from the beginning.

Identification of Significant Variables

As we encounter through observation a totally new phenomenon, our natural
first question is—What causes this? The question was asked with respect to
the diffraction of light, radioactivity, superconductivity, pulsars, and every
other physical phenomenon. It is still being asked with respect to the nature
of elementary particles, climatic change, cancer, and many other topics.
Asking questions about causes can lead to philosophic difficulties, and it is
better to recognize that our natural questions about causes and explanations
for phenomena are really questions about the relationships between observed
variables. The flow of electrical current through a conductor, for example,
can be observed by using ordinary lab equipment to depend strongly on the
potential difference across it and not at all on whether the conductor is oriented North–South or East–West, and this observation can be used to guide
future study. This may seem like a foolishly oversimplified example, but at

the frontiers of scientific work when we know nothing about a new phenomenon, we may have to consider a wide range of possibilities. Normally,
the first phase of research on a totally new phenomenon consists of a search
for the variables that seem to be related. By identifying these significant
variables, we narrow the field of investigation to practical levels and facilitate
continued work at both experimental and theoretical levels.

It is interesting that, at this primary stage of scientific development, we
can make relatively definite statements because we are talking about actual
observations. This accounts for the reputation of scientific activities that they
lead to "scientific truth" about the universe. The claim must be restricted to
the early, diagnostic stage at which we identify the significant variables.
Following this come later stages in which we deal with a totally different
type of activity that involves a much lower level of certainty.

Concept of a Model

After we have analyzed a new phenomenon and are aware of the significant
variables, we can proceed to the next level of sophistication. To illustrate this
stage, consider an elementary example. Suppose we were going to paint a
wall and wished to know the amount of paint to order. We would have to
know the area of the wall, so what would we do? The natural reaction would
be to measure the length of the wall and its height and then multiply the two
numbers together. But what would that give us? And why would we think
that the numerical product has anything to do with the wall? When we multiply these two numbers together, we do obtain something, but it is the area
of the completely imaginary rectangle that is defined by the two lengths. This
imaginary rectangle may or may not have any relationship to the wall. The
important thing to notice is that we are dealing with two completely different
categories. First, there is the real wall whose area we need. Second, there is a
completely invented, conceptual rectangle that is constructed from definitions, exists in our imagination only, and in the present case is specified by
the two measured lengths. We are commonly insensitive to this important
distinction because we are all so familiar with the concept of rectangles that a
simple, almost subconscious, glance at the wall reassures us that a rectangle
is a satisfactory representation of the wall.

But suppose we were not able to make that judgment. Suppose we were
blind and had done nothing more than measure the base and one side of the wall
without thinking about angles or any other property of the wall. We could
multiply our two dimensions to obtain an area that had no relevance at all to the
wall if the wall happened to have the shape of a parallelogram. To avoid that kind
of error, as blind experimenters, we would have to recognize the necessity to

check that the imaginary rectangle defined by the two dimensions compared sufficiently closely with the actual wall. To do this, we would have to know the various properties of rectangles, and we would have to compare with the actual wall as many of these properties as possible. For example, we could test such properties as straightness of sides, right-angle corners, equality of diagonals, and so on. Only after a sufficient number of properties had been compared between the rectangular construct and the real wall, and found to correspond adequately, could we have faith that the area of the imaginary rectangle was a good enough approximation to the actual area of the real wall.

The distinction we have been discussing is most significant and must be borne clearly in mind as we pursue scientific work. In all areas of scientific study we shall find, on the one hand, the real world and our perceptions of it, and, on the other hand, hypothetical, imaginary constructs fabricated out of sets of definitions. Such a construct is often called a **model** of the situation, and the use of models is almost universal in our thinking, whether scientific or nonscientific. The painter contemplating the task of painting the wall has in mind the imaginary rectangle. In addition to the real flower that is being studied a botanist is aware of the concept of the particular species to which the flower is assigned. In contrast to the real flower, the species is a construct that has been defined by a standardized list of properties. Economists studying the economy of a country construct models that consist of a set of definitions and equations and have properties, they hope, that are similar to the actual properties of the real economy. As shorthand descriptions of systems, models give us a framework for thought and communication, a basis for calculation, a guide for future study, and many other advantages.

The use of models is universal in scientific work. Models come in many different kinds and they serve many different purposes, but we must remember their most important characteristic—they *all* are invented concepts. They are constructed so that their properties correspond as closely as possible to those of the real world, but no model can ever be an exact replica of its real counterpart. Models belong to different categories; a wall cannot actually *be* a rectangle, nor a wheel a circle. The properties of a model, however, may be *similar* to the properties of the real world, and in general terms the usefulness of a model depends on the extent that its properties do correspond with those of the real world.

Comparison Between Models and the Real World

At the beginning of an experimental study, we are usually unaware of the extent to which the properties of our model and its real-world counterpart correspond. It is necessary, as a basis for all later work, to start by testing the model against the real system. Only if the properties of the model are shown experimentally to be adequately in correspondence with those of the real system are we justified, like the painter about to order paint, in proceeding to the next step.

Notice that, to be useful scientifically, a model or concept must be actually testable against observation. Thus, a proposition regarding the number of angels who can dance on the head of a pin cannot qualify as science. This is not to say that the only useful ideas are those that can be tested against experience, only that other propositions do not come under the heading of science. Those other propositions may perhaps be perfectly valid as mathematical or philosophical statements, or as aesthetic or ethical judgments.

Refinement of Models

In general, an experimental situation contains, first, the system itself, and, second, a model or models of the system. Whatever else is involved, it is an essential part of the experimenter's task to test the properties of the model against the properties of the real system. In principle, our model will inevitably be incomplete and inaccurate. For example, let us return to the problem of ordering paint for the wall. If as blind painters we test the properties of rectangles with increasing precision against those of the actual wall, we inevitably reach a point at which we begin to find discrepancies. If at that point a simple rectangle is not a good enough model of the wall at the new level of precision, we must modify the model in an effort to improve the match between the model and the wall. To do this, we could progressively make small changes in angles or lengths and hope that the areas calculated for the revised models will provide increasingly accurate estimates for the actual area of the real wall. Even with these adjustments, the model remains an invented concept, and the area calculated from the model belongs to the model and not to the wall.

In scientific work generally, we should feel free to change our models at any time as the need arises. The model is our construct to begin with, and it is only an idea that exists in our heads. In contemplating change, our only consideration is the basic usefulness of the idea and its improved utility if it is altered in any way. Because it is presumably impossible to construct for a piece of the natural world a verbal or mathematical description that is the exact and total equivalent of the real thing, a process of continued refinement and eventual replacement of models must be accepted as the natural course of events. It is the normal business of scientists, whether "pure," technological, or social, to use the process of comparing models and systems in a continuous search for improvement in models. This is usually not an easy process. The models we have now are as good as generations of intelligent and hard-working scientists in the past have been able to make them. We should consider ourselves fortunate in our professional work if we are able to make a few small improvements to existing models. Major revisions or the introduction of completely new models are rare and tend to be associated with Nobel prizes.

On the other hand, we need not be totally preoccupied with improving models. Even if no model can be the exact equivalent of the real thing, the properties of our models and systems can frequently correspond sufficiently well for our purposes. If so, we need not be excessively concerned with the remaining defects. We can proceed confidently with our particular task, provided that we remember periodically to recheck the situation and confirm the continued suitability of the model. It is not appropriate to think about "rightness" or "wrongness" of models. We cannot claim that a model is "correct," only that it is "adequate," or "suitable," or "appropriate" for the purposes in hand.

Model Building in the History of Science

It is possible to gain the impression from the foregoing discussion that in scientific development there is some kind of unique sequence that starts with observation and ends with a satisfactory model. Indeed, scientific thinking has quite frequently progressed in this way, but the sequence is not invariable. There are many examples of a basic, invented idea, the foundation of a model, that was the fruit of pure speculation by the originator, without awareness of the observations that could be directly associated with the conjecture. We can recall, as examples, de Broglie's speculation on the wave model of matter, which was published in 1924 before any of the relevant phenomena were observed directly, and Fermi's invention of the concept of the neutrino almost 40 years before the particle itself was directly observed. There is no single process of scientific development, no single "scientific method." Ideas and observations tend to shuffle forward roughly together but with no automatic leadership from one or the other. Regardless of the precise order of development, one point remains invariable—the fundamental activity in scientific experimenting is to compare the properties of models with the corresponding properties of the real world.

We have not discussed at all the processes by which new ideas are introduced to serve as a basis for totally new theories. Sometimes an existing idea can undergo a process of continued refinement and attain closer and closer correspondence with observation without any alteration of the basic concepts on which the theory is founded (for example, the Ptolemaic theory of planetary epicycles). On the other hand, a theory such as Einstein's theory of general relativity or Schroedinger's wave mechanics can be introduced only after completely radical revision of basic concepts and ways of thinking—not a simple process. The manner in which such major revolutions in scientific thinking have occurred is described in the books by Kuhn, Cohen, and Harré that are listed in the Bibliography.

One might think that, following such major revolutions, a superseded model or theory would be immediately discarded to make way for its successor. Indeed, many models or theories have found no lasting usefulness—one does not hear too much these days about phlogiston, or about earth, fire, air, and water—but this is not always the case. Superseded models have quite often sufficiently close correspondence with the system that, usually on account of simplicity, they continue to be very useful. If one wishes to determine the depth of a well by dropping a stone into the water, one does not need to use Einstein's general theory of relativity as a model for gravitational acceleration. The more sophisticated model must be used when circumstances demand it—for example, when we wish to predict the motion of the planet Mercury.

Detailed Comparison Between Models and Systems

To summarize our development so far, we have four ingredients in the scientific recipe: (1) observation, (2) an idea constructed in our imagination, (3) the process of comparing the properties of the idea with the those of the real world, and (4) the possibility of modifying the idea progressively to improve the fit between the model and the system. We now turn our attention to the actual procedures by which we can compare the properties of models and systems. It is not sufficient to have a vague pictorial concept of the situation; to supply an adequate basis for comparison we must be as explicit as possible. This normally requires quantitative observation of the system and mathematical procedures for specifying the model. Let us consider some specific examples and investigate the various levels of sophistication in the methods for constructing models and comparing them with real systems.

Consider an elastic band, suspended from its upper end, from the lower end of which we can hang weights. The most primitive form of construct with respect to the properties of the system would be a verbal description of its behavior. We could say something like: As I hang more weights at the bottom of the elastic band, it stretches farther. This verbal description could prompt us to invent the general concept "springiness" to serve as a model. But if we wish to refine the model to make it more useful for detailed comparison with observation, our purely verbal methods of description start to fail us; we cannot refine such a vague concept as springiness without resorting to the precision of description that is available in numerical and mathematical modes of expression. We would then make a series of measurements of the extension of the elastic band as a function of load, hoping that they will suggest a more explicit concept. We would obtain a set of measurements such as those shown in Table 4–1.

TABLE 4–1 Extension Versus Load for a Rubber Sample

Load, kg	Extension, m
0.05	0.03 ± 0.01
0.10	0.04
0.15	0.08
0.20	0.13
0.25	0.19
0.30	0.30
0.35	0.34
0.40	0.38
0.45	0.39

(Notice that for simplification we are pretending to know the weights exactly so that we can ignore the uncertainty in them. The values of extension for the rubber band are measurements made by us, and so the uncertainty must be included.)

Now that we have the measurements, do they give us a complete and adequate description of the results? Not really. It is difficult to judge the behavior of a system from a set of numbers in a table; some form of visual presentation is much superior. A simple graph of the observations can comprise all the information contained in the table and can in addition confer the enormous benefit of facilitating visual judgment of the results. Such a diagram is shown in Figure 4–1, in which we have plotted, in addition to to the central values of the measured variables, the actual intervals over which the measurements of extension are uncertain.

In Figure 4–1 we have done nothing more than plot the observations on the graph. At this stage the set of observations is the only thing we have, and there is no justification for putting anything else on the graph.

This completes the first stage of the process, that is, observation. We must now undertake the next stage, in which we construct a model, or models, of the system.

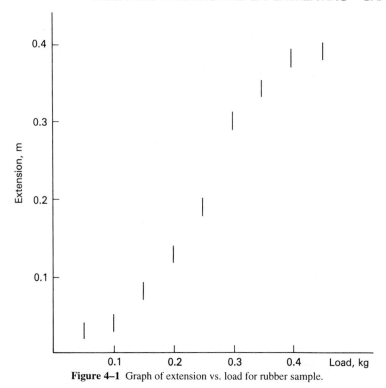

Figure 4–1 Graph of extension vs. load for rubber sample.

4–2 CONSTRUCTION OF MODELS

The type of process required at the various stages depends much on the particular experiment. For example, we may be experimenting on a phenomenon that is being observed for the first time and for which there are no existing ideas. In such a case, our task would be to identify the significant variables and possibly to generate some kind of model. Or we may be working on some relatively familiar phenomenon, in which case we would probably have some existing proposal or theory that could be applied to our system, thus creating a model. Whatever the circumstances, we draw a distinction between models that are **empirical** and models that are **theoretical**. The word *empirical* means that models of this type are based solely on the observations themselves, without any reference to the detailed, internal operation of the system. The processes by which we can generate empirical models and the usefulness of such models is described as we proceed. A theoretical model is constructed more generally, not just for one particular set of observations, and is based on some basic concept or principle about the actual mode of operation of the system. The nature of theoretical models and their usefulness is also described. We consider each type in turn.

Empirical Models

Assume that we have made a set of observations on a system for which there is no existing model. All we have is a set of observations on some property of the system. It could be the load versus extension measurements on our elastic band, and the results probably take the form of a graph like that in Figure 4–1. Our problem is to construct a suitable model. What can we do? There are several possibilities, and we consider them in order of increasing sophistication.

Verbal statement.   The simplest model of all is a simple verbal description of the variation. We could say something like: The extension increases smoothly with load in an S-shaped curve. Notice that even this simple sentence is a construct. As soon as we stop talking about the individual observations and start talking about the whole variation of extension with load, we have made the transition from statements about particular observations to constructed presumptions about the behavior of the system. Even such a vague proposition as the foregoing statement could, on closer measurement, turn out to be unsatisfactory. Perhaps, for example, the variation is really stepwise rather than smooth. The constructed nature of even such simple statements is stressed here as a reminder that we must always be aware of the distinction between statements about the observations themselves and statements that sound as if they were about the observations but are actually statements about our *ideas* concerning the observations.

Drawing a smooth curve through the points.   The next stage of sophistication in model construction is represented by a process that is so commonly carried out (usually without due regard to its significance) that its name is used as the heading for this section. As we view the graph of observations initially (Figure 4–1), we must remember that it contains the observational points *and nothing else*; we have no basis yet for putting anything else on the diagram. There will inevitably be some scatter in the points because of their inherent uncertainty, but it is possible to base the model construction on the single basic assumption that, uncertainty and scatter notwithstanding, the actual behavior of the system is smooth and continuous. This is *our* concept, or idea, and as we draw a smooth curve (Figure 4–2) through the points, we assume that it is valid to apply that concept to our system.

The assumption of smooth, continuous behavior can be valid to a high degree of accuracy for many systems. An example is planetary motion, for

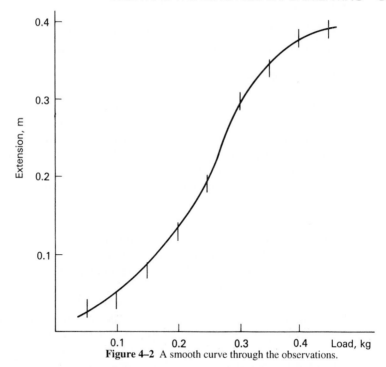

Figure 4–2 A smooth curve through the observations.

which many of the procedures for treating observations were first devised. The responsibility for deciding to assume smooth and regular behavior lies with the experimenter, who should make the assumption only if familiarity with the system leads to the carefully considered conviction that it is valid.

The benefits of assuming regular behavior and drawing a smooth curve through the points can be substantial. One of the most obvious benefits is associated with interpolation and extrapolation. Consider that we have the set of observations shown originally in Figure 4–1 and that we have drawn a smooth curve through the points as shown in Figure 4–2. Our knowledge of the system is good at the points at which measurements have actually been made but if we want to find the value of the extension at a load intermediate between two of the measured values, we have a problem. We could go back to the apparatus and make the desired measurement, but for many reasons this course of action could be either impossible or undesirable. We are then left with the possibility of only *inferring* the desired value on the basis of the existing measurements. The smooth curve provides one way of doing so, as shown in Figure 4–3. We must remember that the answer obtained by interpolation is an *inferred* value that depends on the decision to draw a particular smooth curve.

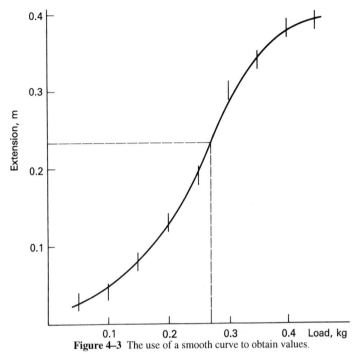

Figure 4–3 The use of a smooth curve to obtain values.

Likewise, it is possible to use a smooth curve to extrapolate *beyond* the existing range of values, as shown in Figure 4–4. Such a procedure enables us to make a guess at values outside the measured range, but the validity of the procedure is obviously much more limited than was the case for interpolation. Before proceeding with extrapolation, we must have very good reasons for believing that the behavior of the system remains regular beyond the measured range. Smooth variation inside the measured range does not by itself offer any guarantees about a wider range of behavior (Figure 4–5).

Mathematical methods for interpolation and extrapolation are given in Appendix 3, and they can be used to obtain interpolated and extrapolated values by calculation, without actually drawing the smooth curve. Such methods still depend on the assumption of smooth, regular behavior of the system, and the inferred values draw their validity from the reliability of that assumption.

Because the validity of interpolation and extrapolation is limited by the assumption of smooth and regular behavior, opportunities for error abound. If, for example, we were to offer someone the graph of temperature versus time (Figure 4–6), without specifying the system, and ask that person to infer the value of temperature for a time halfway between two measured values,

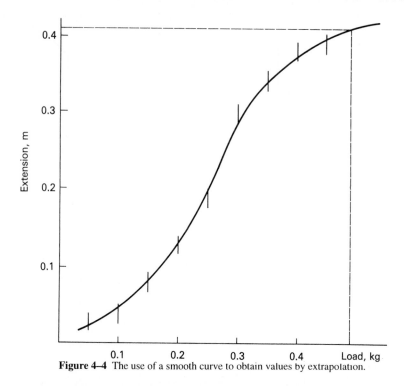

Figure 4–4 The use of a smooth curve to obtain values by extrapolation.

the usual answer would be to draw the smooth curve and obtain the interpolated value as shown on the graph. We could then reveal that the graph depicts the noonday temperatures for the first few days of this month and that we were asking for a temperature at midnight. Likewise, people who have belief in the infallible validity of extrapolation can be asked why they have not made a fortune on the stock market.

Before we leave the topic of drawing smooth curves through points, one final procedure deserves mention. We commonly encounter graphs in which the points have been connected by straight-line segments, as shown in Figure 4–7. Spreadsheet or graphics programs for computers usually automatically draw line graphs in this form. How are we to interpret such a diagram? Surely we are not being asked to believe that these segments represent the actual behavior of the system between the measured points. The only possible benefit seems to be to supply some kind of emphasis. In a diagram containing a number of possibly intersecting graphs, the segments do help sometimes in identifying the various graphs. However, such segments represent satisfactorily neither of the two basic ingredients of experimenting, observations and models, so their use is rarely beneficial and they can be mis-

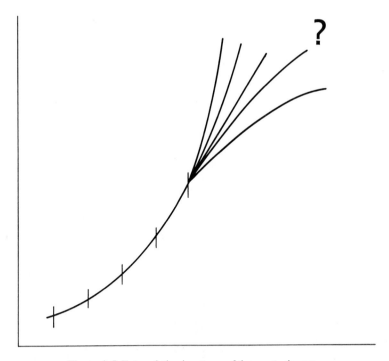

Figure 4–5 Extrapolation is not one of the exact sciences.

leading. For scientific work they are not recommended except in special cases. The way in which the common spreadsheet programs can be adapted for our purposes is described later.

Function finding.   As a more sophisticated form of drawing smooth curves through points, we can use a variety of mathematical methods to find various analytical functions the graphs of which match, to a greater or lesser extent, the variation of the measured values. Obviously, despite all the mathematical sophistication that may be involved, such procedures still depend for their validity on the basic assumption of regular behavior in the system; the curves and functions are *our* concept of the behavior of the system. Nevertheless, functions generated empirically to fit sets of observations can be useful. As mathematical models of the system, they can, be used with varying precision, to obtain inferred values for some characteristic of the system by inter-polation and extrapolation.

It is important to remember clearly that interpolation and extrapolation using an empirical function depend on the validity of that particular function as a model for the system. This does not usually pose too much of a problem for interpolation, where good knowledge of the actual behavior of the system

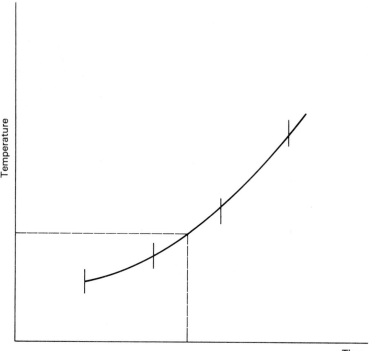

Figure 4–6 Potentially fallacious use of interpolation.

is available on both sides of the interpolated value. Extrapolation is a different matter. We usually notice this when we are dealing with extrapolation in time; forecasting is an uncertain business. We can forecast accurately the time of sunset a week away because the relevant models are very good. We have much less success in forecasting the weather a week ahead because our models are much less satisfactory, and forecasting in other areas, such as the stock market, proves to be almost impossible.

It suffices at present to note the possibility of constructing empirical mathematical models of systems. The methods for doing this are described in Chapter 6.

Theoretical Models

Theoretical models are part of familiar theoretical physics. All analytical theories in physics are constructed out of basic building blocks—definitions, axioms, hypotheses, principles, and so on, followed by analytical derivation from these basic starting points. Because all the elements of theories are constructs of human imagination, the theories themselves and the results of

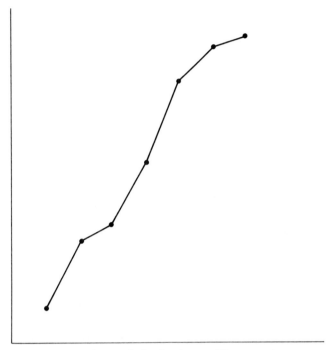

Figure 4–7 A graph showing the use of line segments.

the theories are similarly imaginary constructs. Their relevance to actual systems must be evaluated through experiment.

Let us illustrate the situation by using a particular example. Consider a system in which we can release a steel ball bearing to fall freely under gravity, and we measure its time of fall from various heights. If we wished to construct an empirical model of this system, we could simply measure the time of fall over a number of different distances and graph the result, which would look something like Figure 4–8. We could then use one of the techniques from the preceding section to obtain an empirical model for the system. If we wished to construct a theoretical model of the situation, however, our approach would be completely different. We would have to choose a set of basic axioms or hypotheses from which we would derive the required results. For example, we might decide to use as a basic hypothesis a presumed value for the acceleration of the ball bearing:

$$a = 9.8 \text{ ms}^{-2}$$

Notice that this hypothesis already contains several assumptions about the system, thereby starting our process of constructing an invented model. By

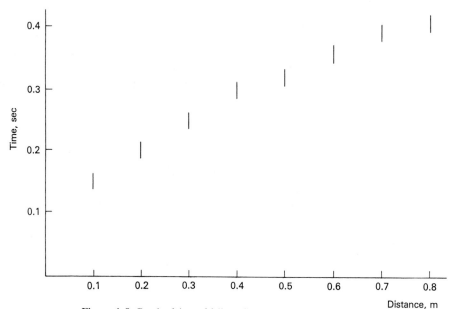

Figure 4–8 Graph of time of fall vs. distance for a freely falling object.

choosing a constant value for the acceleration, we are implicitly neglecting the presence of air resistance. We have every right to do so. The model is ours; we are free to construct it in any way we please. Whether that assumption makes it a *good* model we may not yet be able to tell. As a second example, we are also neglecting effects associated with general relativity; whether this is a serious defect also remains to be seen. We should try to estimate in advance the validity of the assumptions that are built into the model, but we are often limited in our ability to do this. There is always some point at which we have to decide to start experimenting on the basis of the model as it is and to rely on the experimental results to tell us if further refinement of the model is necessary.

We are now ready to proceed with the development of our theoretical model. By integration we obtain

$$v = 9.8t \qquad \text{assuming} \qquad v = 0 \text{ at } t = 0$$

and

$$x = \frac{9.8}{2}t^2 \qquad \text{assuming} \qquad x = 0 \text{ at } t = 0$$

or

$$t = \left(\frac{1}{4.9}\right)^{1/2} x^{1/2}$$

We have chosen to write the final equation in a form in which t is expressed as a function of x because this corresponds to the way in which the experiment was set up. We chose x as the input, or independent, variable and measured t as the dependent, or output, variable. So we want our equation to tell us t as a function of x.

In the course of the derivation, all the assumptions that we insert constitute further components of the model. The final result for the measurable variable, the time of fall, is thus a property of the model. Its applicability to the system is the next topic of investigation.

4–3 TESTING THEORETICAL MODELS

Consider actual measurements for the free-fall experiment. We have treated the distance of fall as the independent, or input, variable for which *we* chose the values; the time of fall is then the dependent, or output, variable of which the system gives us the values. The results of the experiment are shown in Table 4–2. In this experiment, the measurements of the distance of fall could be made much more precisely than those of the time of fall. For simplification, we consider the uncertainty in the x values to be negligible. Normally, we would have to include the uncertainty in all the measured quantities.

TABLE 4–2 Experimentally Measured Time of Fall Versus Distance for a Freely Falling Object

Distance, x, m	Time, t, s
0.1	0.148±0.005
0.2	0.196
0.3	0.244
0.4	0.290
0.5	0.315
0.6	0.352
0.7	0.385
0.8	0.403

The measurements given in Table 4–2 describe the behavior of the system. We also have the behavior of the model, in the form of the function that was the outcome of the analytical derivation:

$$t = \left(\frac{1}{4.9}\right)^{1/2} x^{1/2}$$

The task is somehow to compare these two, but it is not at all clear how that should be done. One simple suggestion is to insert the various values of x in the equation and calculate corresponding values of t. We could then compare these with the measured values. If they agreed exactly, we could perhaps be confident that the system and the model were in correspondence. The probability of that happening, however, is minute; apart from anything else, the presence of uncertainty in the measurements eliminates the possibility of exact correspondence. The major point, though, is that the model is most unlikely to be totally free of systematic defects and deficiencies. It is one of the principal purposes in experimenting to detect these discrepancies and deal with them constructively. The possibility of doing this effectively by using simple arithmetic comparison is small. Much more significant for our purpose is the *overall* behavior of the system; the best way to view that behavior is on a graph.

The graph of our observations, shown in Figure 4–9(a), consists of a series of points. The graph of the model's behavior is a curve, which is shown in Figure 4–9(b). The two graphs together give us a visual impression of the relationship between the properties of the system and those of the model. The comparison would be more detailed yet if we could pick up one of the graphs and lay it over the other. By doing so, we obtain the composite diagram shown in Figure 4–9(c). Notice that this diagram has two different components: (1) points representing the properties of the system, and (2) a line corresponding to the analytical function that belongs to the model.

At last we can make a detailed comparison between the overall properties of the system and those of the model. By straightforward visual inspection, we can say that the model and the system are in correspondence, or are divergent, or whatever. We list the various possibilities in more detail in Chapter 6. For the present, we must note carefully the kind of statement we are able to make at the end of an experiment. We can say only that the behavior of the model and of the system were in correspondence (or were not) to such and such an extent. It is pointless to agonize over whether a theory is "true," "correct," "wrong," or whatever. As was mentioned when we first discussed the nature of models, we should avoid using such terms, even if we are sure we understand the situation. Others may be less clear about the use of words than we are, and there are too many opportunities to be misunderstood. It is far better to categorize a theory or model as "satisfactory" or "good enough," or some similar phrase, because all such decisions are relative to the purposes we have in mind.

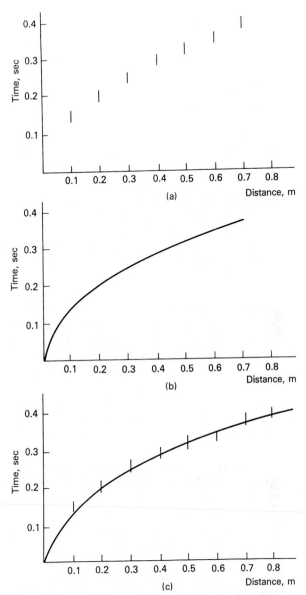

Figure 4–9 The process of comparing the properties of a real system with the properties of a model.

For example, our simple model of constant acceleration under gravity is perfectly satisfactory for finding the depth of a well by dropping a stone down in it, but it is not satisfactory for calculating the trajectory of a space vehicle en route to the moon. If that were our purpose, we would have to construct a more refined theory until we had one that is good enough for that purpose. Even then, we would find that a theory adequate for moon rockets is inadequate to describe the motion of the planet Mercury. For that, as was mentioned earlier, the theory of Newtonian gravitation must be replaced by Einstein's theory of general relativity. And the adequacy of Einstein's theory for describing Mercury's orbit (at a particular level of precision) does not "prove" that it is true or correct, simply that it is good enough for that purpose. Equally, the presence of Einstein's theory does not discredit either Newton's theory of gravitation or our simple model of constant acceleration under gravity. Most people do not measure the depth of wells by using Einstein's relativity theory. In general, we use a particular theory because it is good enough for the purposes in hand. If increased precision is desired at any time, the necessary refinements can be introduced as required (unless, of course, we are working at the limits of knowledge in a particular area, and the chief obstacle is the absence of an improved theory).

Because we are no longer going to use the misleading concept of the "truth" or "correctness" of theories and models, we shall be dependent on our own decision that a chosen model is good enough, or not, for our purpose. One of the primary aims of experiment design is to test the models we use and check their suitability. If it is properly planned, the experiment itself will tell us whether the model or theory is good enough.

In passing, we can note one interesting point of philosophy. Even when our system and model seem to be in complete correspondence, we have to be careful about stating the outcome. All we can say is that, at a particular level of precision, we have failed to observe any discrepancy between the system and the model. It is possible to be more assertive if we are sure that the properties of the model and system are in *disagreement* by an amount clearly in excess of the measurement uncertainty; we can say definitely that the model is not in correspondence with the system. We can say that we have "proved" the theory to be "wrong"—although, even in this case, it would be better to call it "unsuitable" or "inadequate."

Before proceeding, we must note that this process of comparing systems and models depends on our ability to draw the graphs of the functions that appear in the models. At one time this presented substantial difficulties for even simple functions, such as parabolas, and often insuperable difficulties for more complicated functions. Now, the use of computers allows us to make the comparison directly. We can display the graph of our experimental values along with the graph of the function that appears in the model, and judge immediately how well the two correspond. The ways in which common spreadsheet programs can be made to do this are described in Chapter 6.

Even when taking advantage of the marvelous opportunites offered by computers, we must not neglect the development of our own personal expertise in experimenting. First, when no computer is available, we are forced to rely on our own resources. Second, even when we are engaged in computer-based evaluation of experiments, we stand the chance of obtaining completely meaningless results unless we are clearly aware of every detail of the computations that are being carried out invisibly by the computer. To develop our personal expertise, we must turn our attention (as was recommended earlier for calculations on frequency distributions and standard deviations) to the old-fashioned procedures that still constitute the basis of almost all experimenting.

If we are restricted to pencils, graph paper, and rulers, we are virtually compelled to compare the model with the system by using the only function whose graph is easy to draw—a straight line. (For fairly obvious reasons we exclude from consideration that other simple graph, the circle.) A large amount of experimental analysis is still carried out in linear form, and even if the graphical techniques are cumbersome and tedious in comparison with the ease of computer-assisted evaluation, the methods are sufficiently powerful, important, and commonly used that we must become familiar with them.

4–4 USE OF STRAIGHT-LINE ANALYSIS

The objective is to arrange the plotting process so that the behavior of the system and the model are represented on a graph in linear form.

Consider the function for the time of free fall under gravity

$$t = \left(\frac{1}{4.9}\right)^{1/2} x^{1/2}$$

This function, when plotted on an x, t graph has the shape of a parabola. Consequently, if we were to plot the measurements of x and t, intending to compare them with the graph of the function in the model, it would be almost impossible to judge visually whether our results were compatible with a parabola. Suppose, however, we were to plot as variables, not t versus x but t versus $x^{1/2}$. Tthe equation that appears in the model

$$t = 0.4515(x^{1/2})$$

would then take the form of a straight line, whose equation we write

$$\text{vertical variable} = \text{slope} \times \text{horizontal variable}$$

in which the vertical and horizontal variables will be given by

$$\text{vertical variable} = t$$

$$\text{horizontal variable} = x^{1/2}$$

and

$$\text{slope} = 0.4515$$

The experimental values of $x^{1/2}$ and t are given in Table 4–3 and are plotted in Figure 4–10. This graph also contains the line representing the function; the resulting simplification is immediately obvious. The whole process of comparison is facilitated, and we can identify immediately the degree of correspondence between the model and the system.

TABLE 4–3 Experimentally Measured Time of Fall Versus Square Root of Distance for a Freely Falling Object

Distance, x, m	(Distance)$^{1/2}$, $x^{1/2}$, m$^{1/2}$	Time, t, s
0.1	0.316	0.148 ± 0.005
0.2	0.447	0.196
0.3	0.548	0.244
0.4	0.632	0.290
0.5	0.707	0.315
0.6	0.775	0.352
0.7	0.837	0.385
0.8	0.894	0.403

In this example we chose the time of fall as the output variable and the distance of fall as the input variable. This choice caused us to plot the variables as t and $x^{1/2}$. The process would have been equally effective and probably more convenient if we had plotted t^2 versus x instead of t versus $x^{1/2}$. The slope would have been different, but the opportunity to compare the model and the system would have been equally good. In an experiment, there may be several equivalent ways to plot the variables in the form of a straight line. Sometimes one is more convenient than another; sometimes one gives us a better basis for comparing the system and the model. We have to make a decision each time on the basis of the particular circumstances.

Finally, in this example the properties of the model were completely specified, and the line on Figure 4–10 representing the model's behavior is

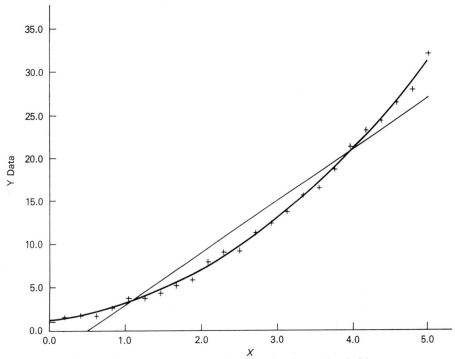

Figure 4–10 Comparison between the properties of a model and of the
system when expressed in straight-line form.

unique. The situation is slightly different if the model contains quantities of
which we do not know the value; this is the topic of the following section.

4–5 CASE OF UNDETERMINED CONSTANTS

Suppose we are doing an experiment on a spring to determine the extension
under various loads (again assuming that only the extension has an uncertainty that needs to be taken into account). Suppose we are aware of a proposal (due to Hooke) that extension x can be considered to be proportional to
load W. This proposal, expressed in mathematical form as

$$x = \text{constant} \times W$$

constitutes an invented model of the system. Assume that we wish to test this
model against the system. The only trouble is that (unlike the former example
of the falling ball in which the model contained the known value of the
gravitational acceleration) we may not know the value of the constant (the

"springiness") that appears in the equation defining the model. Suppose we have made measurements of extension versus load and plotted them in Figure 4–11(a). What are we to do to represent the behavior of the model? The equation

$$x = \text{constant} \times W$$

really represents the infinite set of lines on the $W - x$ plane that have all the values of slope, from zero to infinity, that correspond to the infinite range of possibilities for the value of the constant. Some of these lines are represented in Figure 4–11(b). What, then, constitutes the outcome of our comparison? Laying one graph on top of the other produces the diagram shown in Figure 4–11(c) and provides us with the opportunity to choose a line that is compatible with the experimental points.

But which line or lines are we to choose? Clearly the lines OA and OB have no obvious relevance to the observations and can be disregarded. We can, on the other hand, identify a certain bundle of lines that fall within the region of uncertainty of the measured points. We can estimate visually the edges of this bundle; they are represented by the lines OC and OD. Within these limits all the lines have some degree of consistency with the observations, but no single line stands out as uniquely suitable. All we can say for the moment is that the observations are compatible with the model over a certain *range* of slopes. This means that there is a certain range of values of "springiness" (within the model) that are consistent with the system. The conclusion is, then, that if we have an initially undetermined constant in the model, the experimenting process can be used to determine, within a certain interval, the value that is appropriate for the system. This is the normal way of determining experimental values of physical quantities.

The process is so commonly used because, in addition to the almost necessary use of the graph in comparing models and systems, graphical methods of obtaining values of experimental constants offer so much additional advantage in increased precision that their use is compellingly attractive. The opportunities for error when using algebraic computation alone without graphical checking are great. For example, suppose we are trying to obtain a measured value such as the electrical resistance of a resistor from the variation of the voltage across it with current through it. We make pairs of measurements of I and V, and we use the relationship $V = RI$ directly to obtain a value of R from each pair of I, V values by purely algebraic means. We then hope to obtain an accurate value for R by calculating the average of all the resulting R values.

This approach is deficient in many ways. Basically it fails to satisfy the primary requirement—to compare the properties of the system and the

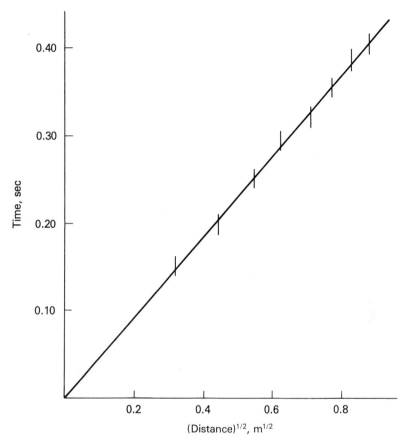

Figure 4–11 The use of straight-line analysis to obtain the value of an undetermined constant in a model.

model—and the consequences for the accuracy of the R value can be serious. If all our pairs of I, V values gave the same, or nearly the same, value for R, we might feel confident in our measurement of R, even without drawing the I, V graph. In the much more likely event that the R values do not all turn out to be the same, we have no way of interpreting the variations without the graph.

We might, for example, encounter a case in which, as plotting the graph would have revealed, the points show more scatter than we expected [see Figure 4–12(a)]. Using a graphical approach, we could still choose a suitable straight line (passing through the origin, if we are sure of the origin as a measured point) and feel reasonably confident about the R value obtained from the slope. Our confidence is justified because the appearance of the graph convinces us that we are dealing with simple scatter about a basically

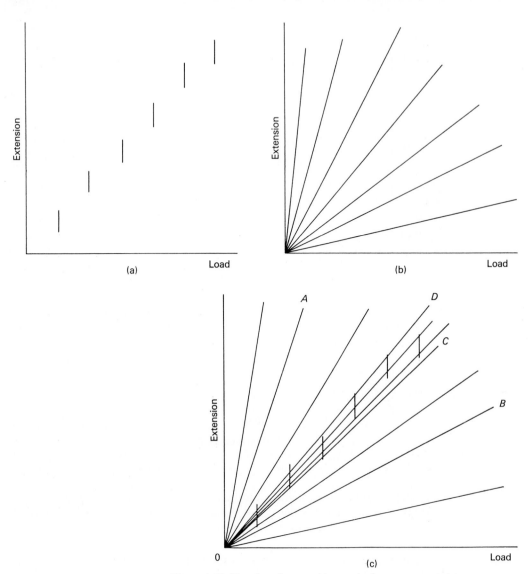

Figure 4–12 Use of graphs to avoid errors in slope measurements.

linear variation. Our nongraphical, algebraic calculation would, on the other hand, give us values that correspond to the slopes of the lines *OA*, *OB*, *OC*, and so on. In a simple table of values, the resulting variability would make no sense at all, and we would gain no insight into what is happening.

As a more significant illustration of the inadequacy of an algebraic, nongraphical approach, consider a case in which some failure of correspondence between the model and the system gives rise either to an unexpected intercept or to deviation from linearity beyond a certain range. Using a graphical approach, we can easily detect and compensate for these discrepancies between model and system. In the first case the graph enables us to obtain a reliable value of R from the slope, which is in such cases unaffected by the presence of the intercept. In the second case we obtain the R value from the slope of the linear portion of the I, V variation, rightly dismissing the nonlinear points as lying outside the scope of the model. The opportunity to make these judgments can come only from visual inspection of the graph, which makes the situation clear at a glance. As before, nongraphical, algebraic calculation from the pairs of I, V values alone yields values of R corresponding [see Figures 4–12(b) and (c)] to the slopes of OA, OB, OC, and so on. These slopes have nothing to do with the slope we want; if we were to include them in an average of algebraically calculated quantities, we would succeed only in introducing error into our answers.

As we ensure in these ways that the final answer is free from such sources of systematic error, it does not matter whether or not we know the origin of the discrepancy. For the purposes of obtaining the value of the quantity under study, it is sufficient at this stage merely to identify the existence of the discrepancy and to ensure that it is not permitted to introduce errors into the answer. We can later consider possible sources of the discrepancy.

We have discussed the process of obtaining values for initially undetermined constants in terms of slopes only. In principle, since a line has two degrees of freedom on a plane, it is possible to obtain from it two pieces of information independently, such as a slope and an intercept. Because of this, an experiment can be made to yield values for two separate quantities that appear in a model. Specific methods of doing this in actual practice are discussed in Chapter 6.

5

Experiment Design

In Chapter 4 we described the various circumstances in which we compare the properties of models and of systems. We encountered such variety that it will come as no surprise to learn that there is no single way to plan experiments. The techniques and procedures we use depend on circumstances, and we describe procedures that are appropriate to a number of cases. The list is not exhaustive but it identifies general principles that are valid in a wide variety of experimental circumstances. Foremost among these is the following. Whatever the circumstances we encounter, we make sense in our experiment procedures only if we keep clearly in mind this central point—the fundamental requirement in experimenting, whatever else is going on, is to compare the properties of a system with the properties of a model or models.

Assume at the outset that, as the outcome of preliminary investigation, we already know the significant variables. Some of these are under our control and can serve as input variables. Others take values determined by the system they are the output variables. In the following sections, we assume that the input variables can be separated and individually controlled; otherwise, with everything varying simultaneously, the interpretation of the results is much more difficult. This unfortunate circumstance is frequently encountered in professional experimenting, but we restrict ourselves here to the case of a fully controlled experiment.

5–1 TO TEST AN EXISTING MODEL

In this section we are concerned with situations in which a model of some type is already available. This model can be the simplest of suggestions (whether theoretically derived or empirical), such as $F = kx$ or $V = RI$, or it can be derived from some grand, sophisticated theory such as Einstein's theory of general relativity. Whatever the nature of the model, its properties almost invariably will take the form of a functional relation between two or more variables. The primary objective is, as always, to compare the properties of the model with those of the system. Only after we have by experiment, satisfied ourselves that, over some range at least, the properties of the system and of the model overlap, are we entitled to go ahead with the evaluation of the quantity we wish to measure.

Notice that any decision about the appropriateness of the model for the system must be based on the experiment itself. We are, of course, not going to attempt to decide on such meaningless questions as whether the model or theory is "true" or "false," "correct" or "incorrect," or whatever. As we have said so often, all models are imperfect in principle, and we simply need to know if the model is good enough for *our* purposes at our level of precision. Only our own experiment can provide the basis for making that decision, and it is one of our major objectives in designing the experiment to ensure that this will be possible. Once we have checked that our model is good enough, we can proceed to evaluate our unknown quantity, not forgetting that, if our situation changes and increased precision is called for, we must reopen the question of the model's adequacy for our purpose. As has been described in Section 4-3, almost invariably the best ways of testing models of physical systems involve a graphical approach. In principle we wish to draw a graph of the model's behavior and to superimpose on it our observations of the behavior of the system. To do this in simple form, however, there are some requirements.

First, because a graph (as we are considering it) is a two-dimensional diagram, we must limit ourselves initially to two variables. In many cases, this requirement is automatically satisfied, as it has been in all the preceding examples. In others, however, the output variable is a function of two (or more) independent input variables. We cannot plot three variables as coordinates on a two-dimensional piece of graph paper (although three-dimensional diagrams can easily be generated by computers and are frequently seen in the scientific literature). Consequently, for our purposes, it is necessary to simplify the experiment by holding one of the input variables constant while studying the dependence of the output variable on the other. We can then alter the first variable to a second fixed value and repeat the process. By a succession of such measurements, we can build up a relatively complete picture of the behavior of the system. Notice that the success of the process depends on the primary assumption that it is possible to hold one of the input variables constant, independently of variation in the other. If such isolation of the input variables is not possible, we have problems; some of the necessary techniques are mentioned in Section 5–2.

For our present purpose suppose that we have only one input variable, either because only one exists or because we can isolate one by holding the others constant. The procedure is clear; we must measure the variation of the output variable with the input variable and plot the resulting values for comparison with the corresponding graph for the model. As was suggested in Section 4-3, however, it would take a computer to draw even simple nonlinear functions, and the advantages of drawing graphs in straight-line form are so overwhelming that we consider only this approach.

5–2 STRAIGHT-LINE FORM FOR EQUATIONS

Simple Cases

If the model we are considering contains only linear functions (such as distance traveled at constant velocity as a function of time, or the potential difference across a constant resistor as a function of current), we have no problem; the equation is already in straight-line form. This is rarely the case, however, and we are almost invariably faced with the necessity of converting the functions found in the model into linear form. We have already encountered this requirement in Section 4-3. There the function was

$$t = 0.4515x^{1/2} \qquad \text{(in units of meters and seconds)}$$

and, clearly, if we wish to represent this equation in the linear form,

$$\text{vertical variable} = \text{slope} \times \text{horizontal variable} + \text{constant}$$

we must choose

$$\text{vertical variable} = t$$

$$\text{horizontal variable} = x^{1/2}$$

$$\text{slope} = 0.4515$$

and

$$\text{constant} = 0$$

This is a simple case, and it is often less easy to see how an equation can be converted into linear form. There are no definite rules for doing it. The best way is to keep clearly in mind the form toward which we wish to work,

vertical variable = slope × horizontal variable + constant

and juggle the quantities in the original equation around until we have the required form. Opportunities for practice are found in the problems at the end of this chapter.

Notice that there is no unique answer in this process. A given equation can sometimes be put in linear form in several different ways. For example, the equation

$$t = 0.4515x^{1/2}$$

can be used equally effectively in any one of the equivalent forms

$$x^{1/2} = \frac{1}{0.4515}t, \qquad\qquad t^2 = 0.2039x, \qquad\qquad x = 4.905t^2$$

with appropriate choices for *vertical variable*, *horizontal variable*, and *slope*. There is a conventional tendency to plot graphs with the input variable horizontally and the output variable vertically, but there is no real requirement to do so. We should choose the form of graph that most effectively serves our purpose.

Our purpose should include not only the basic experimental requirements, but also the comfort and convenience of the experimenter. For this, one should plot variables as simply as possible. For example, consider an experiment to determine a coefficient of viscosity by studying the flow of liquid through a pipe. The appropriate equation (Poiseuille's equation) is

$$Q = \frac{P\pi a^4}{8\eta\ell}$$

where Q = rate of flow
        P = pressure difference across the pipe
        a = pipe radius
        ℓ = pipe length
        η = coefficient of viscosity

In this case the measured variables are P and Q; a and ℓ have constant measured values. Our intention to plot the quantities to obtain a value for η.

One possible choice is to plot Q versus $(\pi a^4 / 8\ell)P$. This seems to suit our purposes because the resulting graph has a slope equal to $1/\eta$, but it is an unwise choice for a number of reasons. First, it greatly increases the amount of arithmetic required to do the plotting, because each value of P must be multiplied by $\pi a^4 / 8\ell$. Second, each of the quantities a and ℓ has an attached uncertainty; if this were to be combined each time with the uncertainty in P, we would have a falsely enhanced uncertainty for the compound quantity (for example, in the actual experiment a would be measured only once, and its uncertainty should not be combined with that of P as if every time we measured P another measurement of a were made to combine with it). Clearly, in this case the path of wisdom is to plot Q versus P and to use $\pi a^4 / 8\eta\ell$ as the slope, thus avoiding all the difficulties just mentioned. We are then able to calculate η from the value for the slope and the measured values of a and ℓ using

$$\eta = \frac{\pi a^4}{8\ell \times \text{slope}}$$

In general, it is best to plot variables that are as simple as possible and to leave most of the arithmetic to be done just once in calculating the answer from the slope.

Use of Compound Variables

In many cases it may suit our convenience (or else it may be absolutely necessary) to plot the graph using variables that are not single measured quantities but are constructed out of the primary measurements. Consider, for example, the so-called compound pendulum, a rigid lamina of a certain shape that is allowed to oscillate under gravity about an axis perpendicular to the plane of the lamina, as shown in Figure 5–1(a). The normal model of its oscillation (for small angles of oscillation) gives the period of oscillation, T, as

$$T = 2\pi\sqrt{\frac{h^2 + k^2}{gh}}$$

where T = period of oscillation (output variable)
h = distance from center of mass to point of support (input variable)
g = gravitational acceleration (constant and unknown)
k = radius of gyration about center of mass (constant and unknown)

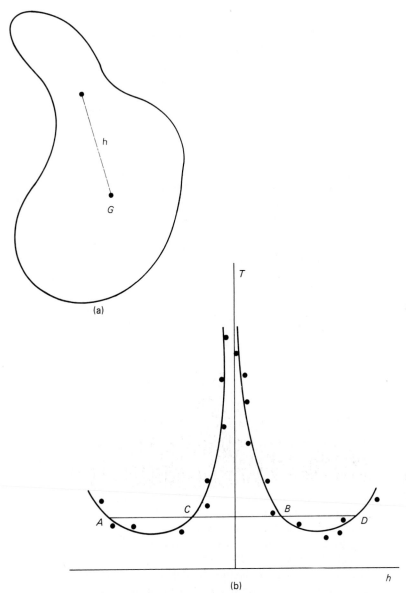

Figure 5–1 The compound pendulum and its variation of oscillation period with h.

Straight-line forms of this equation are not immediately obvious, but it is clearly impossible to place it in the required linear form if we choose the horizontal and vertical variables to be functions of h and T singly. Conversion into linear form by using compound variables is possible. Squaring both sides of the equation, we obtain

$$T^2 = \frac{4\pi^2(h^2 + k^2)}{gh}$$

Therefore

$$T^2 h = \frac{4\pi^2(h^2 + k^2)}{g}$$

and

$$h^2 = \frac{g}{4\pi^2}T^2h - k^2$$

which is now in linear form with

$$\text{vertical variable} = h^2$$

$$\text{horizontal variable} = T^2h$$

$$\text{slope} = \frac{g}{4\pi^2}$$

and

$$\text{intercept} = -k^2$$

Notice how in this case plotting in straight-line form is possible only by using compound variables.

This example is worth studying because it illustrates very clearly the superiority of linear analysis over other methods. A commonly encountered approach to this experiment uses the graph of T versus h, which is shown in Figure 5–1(b). It turns out that the graph directly supplies only the value of k, which can be obtained from the lengths of the intercepts AB and CD. If g is required, it must be obtained by calculation from the value of k. The advantages of the linear form of analysis are clear. First, the T versus h graph gives no basis for comparing the system with the model, unless one uses a computer to draw the graph of the function $T(h)$. Second, no reliable estimate of the uncertainty of the final answer can be obtained from this graph, whereas the overall uncertainty can readily be obtained from the linear graph. Third, the use of an intercept at such a low angle, as illustrated in Figure 5–1(b), is unreliable, because small changes in placing the lines can cause large changes in the length of the intercepted portion; a linear form, on the other hand, enables us to determine the slope of the graph reliably. Fourth, by using the intercept method the answer is determined solely by a few points in the vicinity of the intercepts, and we obtain no benefit from all the other points. When drawing a straight line, however, all the points can contribute to the choice of the line. Last, the linear graph gives g and k from almost independent measurements on the graph, whereas with the other method any inaccuracy in the value of k is propagated automatically into the value of g.

The use of compound variables can also be convenient when there are two or more separate input variables. In such cases, even if the use of compound variables is not absolutely necessary for linear plotting (as it was for the compound pendulum), they often provide the neatest and most effective way of plotting graphs. It was mentioned earlier that, if a system involves two independent input variables, we can study the variation of the output variable with either input variable in isolation, while holding the other input variable at a number of discrete levels. For example, if we wish to measure the specific heat of a fluid, C, by continuous-flow calorimetry, we can allow it to flow at a certain mass flow rate, m, through an electrically heated tube in which the rate of heat generation is Q per unit time. The equation for the resulting heat balance (neglecting losses, etc.) is

$$Q = mC\Delta T$$

where ΔT is the difference in temperature between the input and output ends of the tube. Clearly, both Q and m are separately controllable, and we can perform our experiment by studying the variation of ΔT with m, holding Q at various fixed levels, or we can study the variation of ΔT with Q, holding m at various fixed levels. We would then be able to plot either ΔT versus $1/m$, in which case the various slopes would have values Q/C, or else ΔT versus Q, which would give the various slopes as Cm. Neither possibility by itself provides a complete description of the behavior of the system. Another possibility exists. If we treat the product $m\Delta T$ as *one* variable and plot it against Q, we obtain a single graph that summarizes all the information about the system by incorporating both input variables simultaneously, whether we did or did not control the values of m. The slope would simply have the value C, and we would have a neat way of testing the model and obtaining our unknown in one simple step.

Such use of compound variables is common, and, as before, the choice of combination for the variables and the mode of plotting can be made to suit both the convenience of the experimenter and the requirements of the experiment. If any difficulty in interpreting observations appears when they are plotted by using compound variables (unexpected scatter, perhaps, or some systematic deviation from linearity), we can always obtain extra information about the system by reverting to a plot of the pairs of variables individually rather than in combination. By isolating the effects of the separate input variables we can usually identify the cause of any difficulty.

Logarithmic Plotting

It is frequently desirable and sometimes absolutely necessary to plot variables in logarithmic form. For example, many physical processes involve exponential functions of the form

$$y = ae^{bx}$$

where y and x are measured variables and a and b are constants whose values are to be obtained from the experiment. The equation can be put in linear form by taking logs of both sides to the base e. We obtain

$$\log_e y = \log_e a + bx$$

Thus, if we plot $\log_e y$ vertically and x horizontally (known as a "semi-log plot"), the model gives us a straight line. The slope gives a value for b, and the intercept gives the value of $\log_e a$. Notice that, if logs are taken to the base 10 instead of e, only the intercept is affected, and this can be convenient if we are interested in the slope only.

The use of such logarithmic plotting is common because of the frequent occurrence of exponential functions in the models of physical and chemical processes. In addition, logarithmic plotting is used even for simple algebraic functions. Consider, for example, the function

$$y = x^2$$

Taking logs of both sides, either to base 10 or base e, we obtain

$$\log y = 2 \log x$$

This equation is linear with

$$\text{vertical variable} = \log y$$

$$\text{horizontal variable} = \log x$$

and

$$\text{slope} = 2$$

Thus, functional dependence like a simple square can be tested by using such logarithmic plotting (known as a "log-log plot").

But what is the advantage of this type of graph over a plot, as we have been recommending all along, of y versus x^2? One obvious answer is that it allows us to plot on one piece of paper of reasonable dimensions variations that are too extensive for traditional plotting. A range of one power of 10 in our observations can be conveniently plotted on simple graph paper, a range extending over a factor of 100 is difficult, and a factor of 1,000 makes satisfactory plotting impossible. For these very large excursions of the variables, only logarithmic plotting allows realistic representation of the results.

A second advantage of logarithmic plotting concerns the power of the function. If the system is behaving in such a way that the function

$$y = x^{1.8}$$

would be a better model than $y = x^2$, that fact would probably elude us if we plotted y versus x^2. We would simply obtain a set of points that deviate from a straight line, and the source of the discrepancy would not be immediately obvious. The log-log plot, on the other hand, would still give us a straight line, and this would tell us that some function involving a power was still a good model. The slope of the line would not be 2, of course, and the improved value of the exponent, 1.8, would be available from the slope of the log-log graph. In Chapter 6, we consider further the uses of log-log plotting for the construction of empirical models involving such powers. It suffices for the moment to note that, at the stage of designing an experiment, the possibility of semi-log or log-log plotting should be kept in mind if either the type of function or the range of variables suggests that either is appropriate.

5–3 PLANNING AND EXPERIMENTS

We now list the actual, practical steps by which we prepare to do the experiment. These may seem tedious to those whose ambition is to get on with the experiment as quickly as possible and worry later about what to do with the results. Indeed, for many of the simple experiments commonly encountered in teaching laboratories, the painstaking care that we are about to recommend may seem pointless and pedantic. But remember that the simple experiments in teaching laboratories are merely simulation in suitably simplified form of the much more complicated and important situations that will be encountered later in real systems. If in an introductory physics laboratory we forget to measure the wire diameter while doing an elasticity experiment, it is probably not going to matter too much. We can return to the laboratory later and recover the offending measurement; even if we do not, the world will not come to an end. But if ten years later we plan some space-based astronomy with a very expensive telescope in a spacecraft and we notice only when our experiment is in orbit that we have omitted to test the optics properly, the consequences can be very serious. Acquire as early as possible the habit of meticulous and painstaking planning of experiments even if for the moment such planning may sometimes seem overmeticulous and superfluous.

The planning steps are as follows:

1. Identify the system and the model.
2. Choose the variables.
3. Rectify the equation.
4. Choose the ranges for the variables.
5. Consider the overall precision of the experiment.

Identify the System and the Model

This step may seem somewhat trivial, but it is sometimes surprisingly difficult to identify what, exactly, is the system we are studying. Obviously, we cannot do an experiment if we are not clear, right at the beginning, about the topic of the experiment. The actual phenomenon under study is often surrounded by so much measuring apparatus that we can lose sight of the fundamentals. If we encounter difficulty in answering the question—What, exactly, is the system under study?—we can try looking for the answer to the equivalent question—what is it whose properties are described by the model?

Similarly, identify clearly the model and the limitations contained within it. In the falling-ball experiment, for example, are we going to worry about air resistance or not? If we decide to ignore the presence of the surrounding air in the system, we are not being irresponsible—we are merely defining one aspect of the model that will be used. Whether this is a *good* decision will be made clear later by the experiment itself. If the behavior of the system turns out to be in correspondence with the behavior of the model at the level of precision we use, we can be satisfied that there would have been no point in wasting time on small effects. If we make a poor choice, the results of the experiment will very quickly inform us of the necessity of reconsidering the matter. So at the outset we decide on the limits of the system and the model, and we proceed to test the situation.

Choose the Variables

Usually one quantity in the experiment is an obvious choice for an output variable. If there is only one input variable, there is no problem. If there are several input variables, try to identify one as the chief independent variable and vary the others in discrete steps.

Rectify the Equation

The equation representing the behavior of the model must now be put into straight-line form, as described in Section 5–2. As we have already mentioned, there is no unique, correct choice for the straight-line form. Choose a form that suits the purposes conveniently and effectively. For example, if the equation contains some unknown quantity the value of which is to be determined by the experiment, it is probably best to use a form for the straight line that puts the unknown into the slope. It is possible to determine unknown quantities from intercepts, but because intercepts can frequently be subject to errors arising from instrument defects or other systematic errors, it is usually preferable to obtain unknowns from slopes. If the equation contains two unknowns, it is probably best to find a form that enables us to obtain one unknown from the slope and the other from an intercept.

Choose the Ranges for the Variables

Before starting the actual measurements, we should make decisions about the ranges over which we hope to make them. It is usually best to plan on a range for the input variable of at least a factor of 10. More is better, and less can often give an unsatisfactory basis for comparing the behavior of systems and models. Obviously, we cannot choose directly the range of the output variables; the system itself will tell us these values, but we must still be careful. There may be instrument limits beyond which damage can occur— elastic limits, overheating of precision resistors, overloading of meters and other instruments, and the like. Carefully made trial measurements will allow us to determine the range of the input variables that will avoid overloading any part of the system. This is the time to consider carefully all aspects of instrument ratings, which can be particularly significant for electrical apparatus. For example, does the resistance box have marked on it the maximum current for each range? If so, we incorporate that limit into the choice of range for the variables. If the limits for some piece of equipment are not marked on the equipment itself, we must find the values in the manufacturer's catalogue. In all cases we must ensure that limits are identified and observed. It is too late when the smell of overheated insulation or a vertical column of blue smoke above a meter alerts us to the frailties of physical apparatus and the expense of replacement.

Consider the Overall Precision of the Experiment

We should not start an experiment until we have a general idea of the precision we hope to attain in the overall result. This is not to say that we can guarantee a final level of precision, but we should have a target figure to serve as a guide for our choice of measurement methods. For example, the request—Measure the acceleration of gravity using a pendulum—is by itself virtually meaningless. In response to this request, we could spend ten minutes with crude apparatus and obtain a result with a precision of 10%, or we could spend weeks with refined, expensive equipment and attain 0.01%. We can obtain a realistic impression of the expectation only from some request such as: Measure g using a simple pendulum with a final uncertainty around 2%, and try not to spend more than two hours on it. The figure 2% and the specified time give a general idea of the kind of measurement we are being asked to make and enable us to make sensible choices for experimental method and care in measuring.

Whether or not it is specified in the requirement we have been given, we should have such a target in mind for every experiment we do, for only then do we have the basis for realistic design of the experiment. It will give us the opportunity to ensure that all our measurements are of sufficient precision to contribute usefully to the final result and that we do not waste time and effort making some measurement with precision far in excess of that required.

To see how such a design could be carried out, we return to the example of the pendulum and the hoped-for value of 2% for the final uncertainty in the g value. We know that the result for g, although it will be obtained graphically, in essence involves measurements of ℓ and T (in the form T^2). If the uncertainty in any measurement of either ℓ or T^2 is in excess of 2%, therefore, there is little chance that it will contribute usefully to a final determination of g within 2%. Suppose, as a first guess, we elect to restrict the uncertainties in each of ℓ and T^2 to fall below 1%. What are the implications for the measurements of ℓ and T? The first step must be to make trial measurements to assess the absolute uncertainty with which we can make measurements of ℓ and T. Once we have determined these uncertainties, we can find the limits on the ranges of the ℓ and T measurements that allow the precision to be acceptable. Assume that with the apparatus available we are fairly sure that we can measure lengths with an absolute uncertainty of ±1 mm. What is the measured length at which this uncertainty corresponds to precision of 1%? If the requirement is

$$\frac{0.1}{\ell} = 0.01 \qquad\qquad (\ell \text{ in cm})$$

then the value of ℓ is given by

$$\ell = 10 \text{ cm}$$

Thus, so long as the measured lengths are greater than 10 cm, the contribution to the overall uncertainty from ℓ is within the acceptable range, and we have identified one limit on the acceptable range of ℓ.

What are the corresponding implications for the measurements of T? If we are going to ascribe a precision of 1% to T^2, we need a relative uncertainty of 0.5% in T. The period of oscillation is determined by timing a specified number of oscillations with some kind of timer or stopwatch, and the choice of measuring procedure is determined by the basic uncertainty in that timing device. Suppose we are using a stopwatch that (as we can determine by actually trying the measurement) allows us to measure the time for a number of oscillations to within ±0.2 s. Notice that this figure of 0.2 s must be the overall uncertainty in the *whole* timing process, not just the uncertainty with which we can read the stopwatch once it has stopped. We must include the complete sequence of judging the pendulum's position, pressing the button, and so on. The resulting uncertainty in the overall timing process may substantially exceed the simple uncertainty in reading the scale, and we shall probably have to determine it for our particular circumstances by trying the measurement several times. In any case, if we have an overall uncertainty in the timing measurement of ±0.2 s, we can calculate the relative uncertainty of any timed interval, t, as $0.2/t$. This is the quantity that we wish to restrict to values below 0.5%, and so the limiting condition is

$$\frac{0.2}{t} = 0.005$$

The limiting value of t is therefore given by

$$t = \frac{0.2}{0.005} = 40 \text{ s}$$

Thus, provided we choose the number of pendulum oscillations so that we are never measuring times less than 40 sec, we have a good basis for hoping that our measurements of oscillation time will all contribute effectively to an overall determination of g within 2%. We cannot guarantee that such an experimental plan will result in a value for g with an uncertainty no greater than 2%; there is always the possibility of unexpected contributions to meas-

urement uncertainty or of unsuspected systematic error. But at least we can
avoid making measurements that stand no chance at all of contributing usefully to the overall result.

At this stage we must consider whether each measurement is going
to be considered in terms of uncertainty derived from personal estimate,
or whether random fluctuation is large enough to require the use of
statistical methods. If the latter, some trial measurements will allow us to
make a preliminary estimate of the variance and so enable us to choose
the sample size that will be required if we wish to attain a certain level of
precision. At this point we must recall the warnings about inaccuracy in
small samples that were given in Section 3–11. In addition, however, we
should remember that attempts to improve precision by increasing the
sample size can be unrewarding. The expression for the standard
deviation of the mean involves $\sqrt{N}$, so that if a trial sample of 10
measurements suggests the desirability of, say, a tenfold increase in
precision, the sample size would have to be increased by a factor of 100.
A sample size of 1,000 may not be practicable, and we would have to
seek some other route to improved precision.

Whether the measurements are statistical in nature or whether they
have simple, estimated uncertainty, it should be possible at this stage to
decide if each measurement in the experiment can be satisfactorily made.
If it appears that some measurement is restricted to an uncertainty in
excess of the design aspirations, we must either obtain a more precise
method of measuring that particular quantity, or if that is not possible, we
must acknowledge that our former target for the overall uncertainty of the
experiment was unrealizable with the apparatus available, and that
revision of the target value is necessary. Also, by assessing the
contribution of each quantity in the experiment to the overall uncertainty,
we can identify any measurement that makes a dominating contribution to
the final result, either because of low intrinsic precision or because of the
way in which it enters into the calculation (e.g., some quantity raised to a
high power, or a quantity that has to be obtained as the difference
between two measured values). Once identified, these measurements can
be given special attention so that their uncertainty can be kept under
control as much as possible.

All the detail described in this section may seem to constitute an unnecessarily exacting approach to a small, simple experiment, but it is wise to
recall again that we are practicing for much bigger, more important experi-

ments in which the consequences of failure to plan properly can be serious and expensive.

Construction of Measurement Program

After choosing the variables, ranges, and precision of measurement, it is best to conclude the design of the experiment by constructing a complete and explicit measurement program. This will normally take the form of a table that includes all the quantities to be measured in the experiment and that also provides space for any computations required for drawing the graphs. A completed measurement program allows the experimenter to concentrate during the course of the experiment on the actual conduct of the experiment. While one is manipulating apparatus and making measurements, there is usually enough to be done without the continuous necessity of deciding what to do next. The measurement program also helps to guard against the accidental omission of some significant measurement that could be overlooked as a consequence of the pressure of actual experimenting.

The complete process of experiment design is illustrated in the description of a sample experiment in Appendix 4.

As we have remarked frequently, all this planning may seem like an unnecessary amount of fuss for a simple experiment. These recommendations, however, represent nothing more than the basic minimum of preparation for any serious experimenting, and no opportunity should be lost for the early formation of careful habits of experiment design and planning. It is important to avoid the temptation to rush ahead with the experiment, leaving until later the task of deciding what to do with the results; it is much more beneficial to acquire the habit of setting aside the time to design and plan an experiment properly before starting the actual measurements.

5-4 EXPERIMENT DESIGN WHEN THERE IS NO EXISTING MODEL

The problem of designing an experiment when there is no model appears when, for example, we are making observations on some phenomenon that is so new that a theoretical model has not yet been constructed, or else on some system that is so complicated (e.g., a complex engineering system or some aspect of national economics) that it will probably never be possible to construct a satisfactory theoretical model for it. If we do not have an existing model to test, our objective in doing experiments on the system can take several forms. We may be motivated by simple curiosity or by a practical need for information about the system. At a higher level, we may be interested in the possibilities of model building. We may be seeking guidance for the construction of a theoretical model, or if this is too difficult, we may wish to obtain measurements to serve as a basis for a purely empirical model of the system. As was mentioned in Chapter 4, even in the absence of detailed, theoretical understanding, empirical models are extremely useful. They can be helpful in systematizing our thoughts about a complex system, and they are usually essential for such mathematical calculations on the system as in-terpolation, extrapolation, forecasting, and so on.

Whatever the motivation, we need a function or graph that provides a good enough fit to the observations. The methods of finding suitable functions are described in Chapter 6. We restrict ourselves for the moment to the question of designing the experiment. In the absence of an existing model, experiment design can be relatively straightforward, especially if we can isolate the input variables so that we can vary one while holding the others at fixed values. Experiment design can consist simply of measuring the output variable over suitable ranges of the input variables to build up as complete a picture of the behavior of the system as possible. If we cannot isolate the input variables, we have problems, and this case is considered in Section 5–7.

Even if there is no existing theory for a phenomenon, it is wise to ac-cept any available hints about functions that might be appropriate to our system and to test these possibilities against the system's behavior. One way of obtaining such suggestions is discussed in the next section.

5–5 DIMENSIONAL ANALYSIS

Even without a complete theory of a physical phenomenon, it is still possible to obtain useful guidance for the performance of an experiment by dimen-sional analysis. The dimensions of a physical (mechanical) quantity are its expression in terms of the elementary quantities of mass, length, and time, denoted by M, L, and T. Thus, velocity has dimensions LT^{-1}, acceleration LT^{-2}, density ML^{-3}, force (equals mass × acceleration) MLT^{-2}, work (equals force × distance) ML^2T^{-2}, and so on.

The principle used in dimensional analysis is based on the requirement that the overall dimensions on the two sides of an equation must match. Thus, if g is known to be related to the length and period of a pendulum, it is obvious that the only way by which the LT^{-2} of the acceleration on the left side can be balanced on the other side is to incorporate the length to the first power (to give the L) and the period squared (to provide T^{-2}). We can thus say immediately that, whatever the final, theoretical form for the equation, it must have the structure

$$g = (\text{dimensionless constant}) \times \left(\frac{\text{length}}{\text{period}^2} \right)$$

The process can give no information about dimensionless quantities (pure numbers such as π, etc.), and so we must always include their possible presence in equations obtained by dimensional analysis.

The general method is as follows. Consider a quantity z that is assumed to be a function of variables x, y, and so on. Write the relation in the form

$$z \propto x^a y^b$$

where a and b represent numerical powers, initially unknown, to which x and y may have to be raised. Then write down the dimensions of the right-hand side in terms of the dimensions of x and y and of the powers a and b. Second, set down the condition that the total power of the dimension M on the right-hand side must be the same as that known for z. Do this also for L and T, obtaining thereby three simultaneous equations that enable us to calculate values for a, b, and so on.

For example, consider the velocity v of transverse waves on a string. We might guess that this velocity is determined by the tension T in the string (not to be confused with the T that appears as the dimensional symbol for time) and the mass per unit length m. Let us write

$$v \propto T^a m^b$$

The appropriate dimensions are;

of v:	LT^{-1}
of T (force):	MLT^{-2}
of m (mass per unit length):	ML^{-1}

Therefore,

$$LT^{-1} = (MLT^{-2})^a (ML^{-1})^b$$
$$= M^{a+b} \times L^{a-b} \times T^{-2a}$$

Thus, by comparing in turn the powers of M, L, and T on the two sides of the equation, we obtain

$$\text{for } M \qquad 0 = a + b$$

$$\text{for } L \qquad 1 = a - b$$

$$\text{for } T \qquad -1 = -2a$$

of which the solutions are obviously

$$a = 1/2, \qquad \text{and} \qquad b = -(1/2)$$

We obtain finally

$$v = (\text{dimensionless constant}) \times \sqrt{\frac{T}{m}}$$

Such a procedure is very valuable, for even in the absence of a detailed, fundamental theory, it provides a prediction regarding the behavior of the system. This can be a starting point for experimental investigation. If the experiment shows consistency between the system's behavior and the model produced by dimensional analysis, we have confirmation of the validity of our original guess regarding variables. If the experiment shows a discrepancy, we must look again at our primary suppositions about the quantities involved in the experiment. Notice that in the foregoing example we obtained three equations for only two unknowns. The situation, therefore, was really overdetermined, and we were fortunate that the equations containing a and b were consistent. Had that not been the case, we would have known immediately that our guess regarding the constituents of v was wrong.

Powerful as this method is, difficulties obviously arise when the quantity under discussion is a function of more than three variables. We then have more than three unknown powers but only three equations from which to determine them. A unique solution is not possible, but a partial solution may be found in terms of combinations of variables.

For example, consider the flow rate Q of fluid of viscosity coefficient η through a tube of radius r and length ℓ under a pressure difference P. All these quantities are clearly significant in determining the flow rate, and so we may suggest a relation

$$Q \propto P^a \ell^b \eta^c r^d$$

The dimensions of the quantities are as follows:

Q (volume per unit time) L^3T^{-1}

P (force per unit area) $MLT^{-2} \times L^{-2} = ML^{-1}T^{-2}$

ℓ (tube length) L

η (viscosity coefficient,
defined as force per
unit area per unit
velocity gradient) $(MLT^{-2})(L^2)^{-1}(LT^{-1} \times L^{-1})^{-1} = ML^{-1}T^{-1}$

r (tube radius) L

Therefore

$$L^3T^{-1} = (ML^{-1}T^{-2})^a \, L^b \, (ML^{-1}T^{-1})^c \, L^d$$

Comparing powers of

M	$0 = a + c$
L	$3 = -a + b - c + d$
T	$-1 = -2a - c$

Here we have four unknowns and only three equations, so that in general a complete solution is not possible. We can obtain part of it, however, for it is obvious that the M and T equations give us

$$a = 1$$
$$c = -1$$

The equation for Q must therefore contain the term P / η. The remaining part of the solution can be written only as

$$b + d = 3$$

If we write this

$$d = 3 - b$$

we can see that Q must contain the product r^3 / r^b. It also contains ℓ^b, so that we can write

$$Q \propto \frac{P}{\eta} \times r^3 \times \left(\frac{\ell}{r}\right)^b$$

Because it is inconceivable that Q should increase with ℓ, if all other quantities are kept constant, it is obvious that b must have a negative value, and we can invert the ℓ / r term to obtain, finally,

$$Q \propto \frac{P}{\eta} \times r^3 \times \left(\frac{r}{\ell}\right)^b$$

The quantity b remains unknown, and this is as far as dimensional analysis can take us toward the complete solution. Even this partial solution, however, could serve as a guide to experimenting in a situation in which no fundamental theory existed. Dimensional analysis can be extended to cover thermal and electrical quantities, but in those cases ambiguities arise and they require special consideration. The appropriate discussion can be found in the standard texts on heat and electricity or in the specialized texts on dimensional analysis.

5–6 DIFFERENCE-TYPE MEASUREMENTS

In all the preceding sections we have assumed that there was a clear and definite relationship between the input and output variables, and that the input variables themselves were readily identifiable and relatively well controlled. We do, however, encounter circumstances in which we are not so fortunate. Perhaps our input variables cannot be clearly isolated, so that, with everything varying at once, it is difficult to identify the effect of each on the output of the system. Or perhaps the system is so complex and subject to so many variable factors that we find it hard to judge whether the effect in which we are interested even exists. Many experimental techniques, mostly of a statistical nature, have been devised for use in such circumstances. Descriptions of these can be found in the texts on statistics listed in the Bibliography. For our present purpose, we restrict ourselves to a brief description of the problems appearing at the various levels of complexity and uncertainty.

Difference-Type Experimenting in the Physical Sciences

Suppose we wish to study some relatively small effect, such as the extension of a hard steel wire under load. Not only is the effect small, but it also is subject to a number of perturbing factors—for example, temperature. If we simply measure, therefore, the extension of a particular wire under a certain load without ensuring temperature stability, we cannot be certain that the extension we measure can be ascribed uniquely to the influence in which we are interested, namely, load. Not only that but if in addition we are actually unable to control the temperature, we shall never be able to be sure about the effect of load on the wire. The solution is a **null-effect** measurement. We study two identical specimens simultaneously, one loaded, the other not, and we measure the *difference* in length between the two specimens. The wire under load shows the behavior of the system as it is tested, while the unloaded wire provides the null effect—that is, the behavior of the system in the *absence* of the load. We can then hope to ascribe the measured difference in length to the influence in which we are interested, the load, and the perturbing influences that affect the two wires equally are prevented from introducing errors into the measurements. We must obviously try to ensure that, as far as possible, the two specimens be identical, be subject to exactly the same influences, such as temperature, and differ in only the one respect—load.

Fortunately, such correspondence is not too hard to achieve if we are talking about steel wires. We can come close to making the situation of the two wires identical by mounting them close together (to minimize temperature differences between them) and by taking other similar precautions. And because we wish the basic properties of the two specimens to be as close to identical as possible, we can simply take one length of wire and cut it in two, making one piece the specimen to be loaded and the other the comparison specimen that will indicate the null effect. Our ability to cut the specimen in half allows us conveniently to perform a great variety of difference types of measurements and to obtain high precision in the detection of small effects that would otherwise be hopelessly obscured by perturbing factors. Such experimenting is common and is encountered over the whole range of physical phenomena.

It is always good experimental practice to check the performance of an experimental system *in the absence of* the influence we are studying as well as in its presence. Sometimes the results are surprising, and we do well to take the advice of Wilson (see the Bibliography) and reflect on the statement: It has been conclusively proved by numerous tests that the beating of drums and gongs during a solar eclipse causes the sun's brightness to return.

Difference-Type Experimenting in the Biological Sciences

In illustrating null-effect measurements using the extension of a loaded steel wire, we have encountered one very convenient aspect. To guarantee the similarity of the experimental specimen and the comparison specimen, we cut the basic specimen in half. In the case of steel wires and other similar materials, that presents no problem, but other systems are not so cooperative.

Suppose we wished to measure the effectiveness of a new drug for a particular type of illness. It would clearly be of little value if we did nothing other than simply administer the drug to a patient suffering from the disease and watch for improvement. There are far too many variable and perturbing factors for us to ascribe confidently any change in the patient's condition to the drug. If we wish to isolate the effect of the drug alone, we should clearly try to design some sort of difference-type experiment in which we observe the null effect as well as the influence of the drug. Such a requirement raises obvious difficulties not encountered when experimenting on steel wires. The reluctance of most human specimens to be cut in half makes it impossible to create a genuine null-effect specimen. We could use a second person as a null-effect specimen, but we would immediately encounter all the variability of response that we had sought to evade by using identical specimens.

Faced with the inevitability of biological variability, our only recourse is to compensate with increased numbers. We abandon attempts to experiment on single specimens and construct an *experimental group*, which we expose to the influence under study, and a *control group*. The control group is constructed to be as closely comparable as possible to the experimental group, differing only in that it does *not* receive the treatment that forms the topic of the research. It will, we hope, be exposed to all the perturbing influences that affect the experimental group, will respond to them in the same way as that group, and will therefore provide the null-effect measurement.

Many refinements may have to be built into this kind of experimenting, because the effects we seek to measure can often be quite small in comparison with all the perturbing influences. For example, to diminish subconscious distortion of the results in medical experimenting on human subjects, it is common to offer the members of a control group a simulation of the real material given to the experimental group (a placebo), and to keep both the experimenters and the subjects in ignorance of the allocation of real and simulated material (the so-called double-blind experiment).

Experiment designs involving an experimental group and a carefully matched control group are virtually universal in biological studies, whether we are trying to measure the possible carcenogenicity of some food dye in large numbers of unfortunate mice or the beneficial effects of musical activities on the academic achievement of elementary school students.

5–7 EXPERIMENTING WITH NO CONTROL OVER INPUT VARIABLES

Sometimes we have to design a process to study some system over which we have no control at all. If this is the case, we have no alternative to simple unmanipulative observation of the system, and our task is to design the observational procedure (perhaps we are not justified in calling it an experiment) to optimize our chances of effective comparison between the properties of the system and those of any model we have in mind. In cases of clear-cut behavior of the system and well-defined models, we may not have too much of a problem. For example, astronomers may suffer the frustration of inability to influence their subject matter, but their system usually functions in a well-defined manner, often permitting extremely accurate measurement. In this way it is not too hard to decide that Einstein's theory of general relativity fits the observations on the orbit of the planet Mercury better than does Newton's theory of gravitation.

In other cases, however, the questions we ask may be harder to answer. For example, has the introduction of a new detail of manufacture altered the quality of a manufactured product? Even when everything in the manufacturing process is kept as nearly constant as possible, observation shows that the product varies from specimen to specimen. Does this variance mask the effect in which we are interested? Without control over our input variables, the study becomes an exercise in sampling procedures, and a whole field of industrial study exists under the title of "quality control." The literature on statistics and statistical experiment design is extensive; some of the texts listed in the Bibliography provide a starting point.

Even industrial processes, with their inherent fluctuations and their lack of input control, pose problems that are simple in comparison with some of the questions to which we seek answers today. Does the addition of fluorides to municipal water supplies improve the condition of peoples' teeth, and does it have other, possibly harmful effects? Do nuclear power stations cause a higher incidence of leukemia in their vicinity? In seeking the answers, we have almost every problem that can face an experimenter. There is little or no control over the input variables, there is wide variation in individual response, the response may be of only a probabilistic nature, there may be long delays in observing a response, there is rarely an opportunity to observe a genuine null effect (we do not normally carry out surveys of sufficient sensitivity before the municipality starts to add fluorides or before the nuclear power station is built), and there is commonly a multitude of confusing extraneous factors. The only thing we can do is to carry out the sampling procedure as carefully as possible. We must obtain an artificial null-effect measurement by constructing an experimental group as large as possible that is under the influence we are studying and a control group that is exempt from that influence but that in every other respect matches the experimental group as closely as possible.

The whole point in this kind of experimenting or survey work lies in the skill and care with which the sampling is done. The effects under study are usually so subtle that, as a consequence of no more than changes in sampling procedure, it is not uncommon for different surveys to provide completely contradictory conclusions. It is not completely unknown that people with special interests in mind can supply results of surveys to "prove" their point, obtaining the result they want by careful control over their sampling procedures. Many of the issues in which scientific matters have a bearing on public policy have this characteristic of uncontrollable input variables, and we should all become as familiar as possible with the procedures used for sampling and significance testing. In this way we may be able to judge as accurately as possible the usually conflicting claims of the protagonists.

When faced with problems of such complexity, we must frequently abandon familiar patterns of thought that have been successfully used in other areas. For example, the word *proof* is legitimately used in many contexts. We can, for example, prove mathematical results as consequences of mathematical principles. The word is also used (perhaps less legitimately) with reference to measurements when the uncertainty level permits. Most reasonable people would accept it as "proved" that the sun is more distant from the earth than the moon (although it would be better to say simply that the distance is measured to be greater). But there are other areas in which we cannot use the word at all. We have all heard claims from some of the interested parties that the evidence linking cigarette smoking to lung cancer is "only statistical" and that harm has not been "proved." This is a common form of argument in such matters of public policy. Situations in which such controversy exists are usually difficult to deal with, partly because the observable effects may appear only in terms of probabilities, and also because of long delays in the appearance of the effects. In such cases the concept of proof must be modified. It is actually replaced by the concept of *correlation*. Correlation studies give results, phrased in terms of probabilities, that differ in character from the clear-cut cause-and-effect relationships with which we are familiar in other experiments. Nevertheless, they can be equally valid for identifying the factors that influence systems. The concept of correlation receives further consideration in Section 6-14.

PROBLEMS

1. A scientist claims that the terminal velocity of fall of a parachutist is dependent on only the mass of the parachutist and the acceleration due to gravity. Is it reasonable setting up an experiment to check this suggestion?

2. The range of a projectile fired with velocity v at angle α to the horizontal may depend on its mass, the velocity, the angle, and the gravitational acceleration. Find the form of the function.

3. The pressure inside a soap bubble is known to depend on the surface tension of the material and the radius of the bubble. What is the nature of the dependence?

4. The period of a torsion pendulum is a function of the rigidity constant (torque/unit angular deflection) of the support and the moment of inertia of the oscillating body. What is the form of the function?

5. The deflection of a beam of circular cross section supported at the ends and loaded in the middle depends on the loading force, the length between the supports, the radius of the beam, and Young's modulus of the material. Deduce the nature of the dependence.

In all the following problems state the variables or combination of variables that should be plotted to check the suggested variation and state how the unknown (slope, intercept, etc.) may be found.

6. The position of a body starting from rest and subject to a uniform acceleration is described by the function

$$s = 0.5at^2$$

s and t are measured variables. Determine a.

7. The fundamental frequency of vibration of a string is given by

$$f = \frac{1}{2\ell}\sqrt{\frac{T}{m}}$$

f, ℓ, and T are measured variables. Determine m.

8. The velocity of outflow of an ideal fluid from a hole in the side of a tank is given by

$$v = \sqrt{\frac{2P}{\rho}}$$

v and P are measured variables. Determine ρ.

9. A conical pendulum has a period given by

$$T = 2\pi \sqrt{\frac{\ell \cos\alpha}{g}}$$

T and α are measured variables, ℓ is fixed and known. Determine g.

10. The deflection of a cantilever beam is expressed by

$$d = \frac{4W\ell^3}{Yab^3}$$

d, W, and ℓ are measured variables, a and b are fixed and known. Determine Y.

11. The capillary rise of a fluid in a tube is given by

$$h = \frac{2\sigma}{\rho g R}$$

h and R are measured variables, ρ and g are fixed and known. Determine σ.

12. The gas law for an ideal gas is

$$pv = RT$$

p and T are measured variables, v is fixed and known. Determine R.

13. The Doppler shift of frequency for a moving source is given by

$$f = f_0 \frac{v}{v - v_0}$$

f and v_0 are measured variables, f_0 is fixed and known. Determine v.

14. The linear expansion of a solid is described by

$$\ell = \ell_0(1 + \alpha \cdot \Delta T)$$

ℓ and ΔT are measured variables, ℓ_0 is constant but unknown. Determine α.

15. The refraction equation is

$$n_1 \sin\theta_1 = n_2 \sin\theta_2$$

θ_1 and θ_2 are measured variables, n_1 is constant and known. Determine n_2.

16. The thin-lens (or mirror) equation can be written

$$\frac{1}{s} + \frac{1}{s'} = \frac{1}{f}$$

s and s' are measured variables. Determine f. There are two ways of plotting this function. Which is better?

17. The resonant frequency of a parallel L-C circuit is given by

$$\omega = \frac{1}{\sqrt{LC}}$$

ω and C are measured variables. Determine L.

18. The force between electrostatic charges is described by

$$F = \frac{1}{4\pi\varepsilon_0}\frac{q_1 q_2}{r^2}$$

F and r are measured variables. q_1 and q_2 are fixed and known. How do you check the form of the function?

19. The force between adjacent current-carrying conductors is described by

$$F = \frac{\mu_0}{4\pi}\frac{i_1 i_2 \ell^2}{r^2}$$

F, i_1, i_2, and r are measured variables, μ_0 and ℓ are constant. How do you check the form of the function?

20. The discharge of a capacitor is described by

$$Q = Q_0 e^{-t/RC}$$

Q and t are measured variables. R is fixed and known. Determine C.

21. The impedance of a series R-C circuit is given by

$$Z = \sqrt{R^2 + \frac{1}{\omega^2 C^2}}$$

Z and ω are measured variables. Determine R and C.

22. The relativistic variation of mass with velocity is

$$m = \frac{m_0}{\sqrt{1 - \dfrac{v^2}{c^2}}}$$

m and v are measured variables. Determine m_0 and c.

23. The wavelengths of the lines in the Balmer series of the hydrogen spectrum are given by

$$\frac{1}{\lambda} = R\left(\frac{1}{4} - \frac{1}{n^2}\right)$$

λ and n are measured variables. Determine R.

6

Experiment Evaluation

6-1 GENERAL APPROACH

Even when we have finished making the measurements in an experiment, an equally significant part of the process still remains—we must evaluate the significance in what has been done. Our objective in doing an experiment is to be able to make some statement, usually about the relationship between some system and a model. It is important to identify clearly the statement we wish to make and to ensure that the statement is as accurate and complete as possible and fully justified by our measurements. The precise way in which we evaluate the experiment as a whole depends on the type of experimenting we have been doing. As described in Chapters 4 and 5, we may have been operating with or without a theoretical model, and our measurements may or may not have been dominated by statistical variance. The procedures we must now follow will vary accordingly.

Before we proceed, we must note two general points. First, we should always remember that experimental results are precious. They have often been obtained from an extensive experimental program involving many people and large amounts of money. At any level of cost, the results may be unique and irretrievable. We should accept the obligation to extract every available bit of information from the observations and to ensure that our final statement is as complete as possible. The second general remark concerns objectivity. It is almost impossible to avoid approaching an experiment without some preconception of what "ought" to happen. We must discipline ourselves to be as objective as we can, and if the outcome of the experiment is

113

different from what we expected or hoped for or is disappointing in some way, we must be prepared to state the result honestly and realistically and to obtain from it the guidance required for future work.

In the teaching laboratory, where it is sometimes difficult to keep our ultimate objectives clearly in mind and easy to forget that the experiments serve to simulate real tasks in the working world, we commonly encounter the mistaken belief that the objective is to reproduce the known values of experimental quantities. If we measure the acceleration of gravity and obtain a value of 9.60 m s^{-2}, our answer is different from the "right" answer and so we are "wrong." The "error" can then be conveniently blamed on the apparatus. Because there are no "right" answers for experimental quantities, the situation really involves the comparison of two measured values for a quantity. Each measured value has its own characteristics, and each has its own range of uncertainty. To assess the significance of a discrepancy between two independently measured values of a quantity is actually a complex and difficult task. It is far better first to develop the ability to make our own measurements as reliably as possible and to assess their range of uncertainty as accurately as possible; we can worry later about comparing our measurements with those of other people.

So when we make measurements on quantities for which we are sure we already know a more precise value, it is best to discipline ourselves to avoid thinking about the more precise, or "standard," value; it is better to acquire experience and build up confidence in our *own* work. This confidence will be necessary later as we undertake professional experimenting, in which we must take responsibility for the experiments and measure things that have never been measured before.

So if we obtain 9.60 m s^{-2} for g, we must be equally aware that the measured uncertainty is ±0.3 m s^{-2} and that the result is not as bad as we might think at first. If we are going to grumble about anything, let it be the ±0.3 m s^{-2}, but we must not feel guilty about it if the experimental apparatus, with normal effort, is not capable of precision better than 3%. We must not be misled by the way in which accepted values for physical quantities are quoted in textbooks. The values are often mentioned rather casually, and the texts rarely make it clear that these numbers represent the outcome of sophisticated work by generations of expert scientists. It is instructive to read the detailed history of such measurements. Excellent accounts of some of them are in the book by Shamos listed in the Bibliography. We should not be too casual about numbers such as these and should not hope to reproduce them exactly in two hours of work in an elementary laboratory.

The main point is to state the result of the experiment honestly and objectively. The experimenter should strive earnestly to maximize the yield of the experiment by making the final answer as reliable as possible and the limits of uncertainty as close as the experiment will permit, but in all cases it is important to be realistic.

6–2 THE STAGES OF EXPERIMENT EVALUATION

The process of evaluating the result of an experiment has several parts. First, we must obtain the values of the basic measurements and their uncertainties. Second, we must assess the degree of correspondence between the properties of the system and of the model. Third, we must calculate the values of whatever property of the system we set out to measure. Last, we must make an estimate of the overall precision of the experiment. Let us consider each of these steps in turn.

Computation of Elementary Quantities

The first step in working out the result of an experiment consists of calculating the elementary quantities of which the experiment is composed. For example, an experiment on a simple pendulum that has the purpose of obtaining a value for g will probably yield, as its input variable, a set of measurements of length ℓ. The output variable will be presented by a set of measurements of the times required for a certain number of oscillations, and from them values of the period T can be calculated. Our present purpose is to compute the values of ℓ and T and their uncertainties; these will be the basis of the subsequent graphical analysis. The choice of procedure here depends on whether we have elected to make a subjective assessment of the uncertainty range of each measurement or have decided that random fluctuation is sufficiently prominent that statistical treatment is desirable.

Estimated Uncertainty

In the case of the simple pendulum, the first variable to consider is ℓ. Here we may have found that measuring the length of the pendulum with a meter stick has enabled us to identify intervals, as described in Section 2–3, within which we are almost certain our values lie. Our experimental results will therefore take the form of a set of values for ℓ in the form: value ± uncertainty. It is conceivable, too, if we have been counting swings and measuring the times with a stopwatch, that we are similarly able to identify intervals on the time scale within which we are almost certain that our values for time lie. These, too, would be expressed as time value ± uncertainty. This, however, is not yet our variable T. We might have counted 15 oscillations of the pendulum, obtaining a time value of 18.4 ± 0.2 s, and the value for the period, the time required for *one* oscillation, must be obtained by division as 1.227 ± 0.013 sec. Notice that not only the central value must be calculated in this way but also the value of the uncertainty. In simple algebraic terms

$$\left(\frac{1}{15}\right)(18.4 \pm 0.2) = \frac{18.4}{15} \pm \frac{0.2}{15} = 1.227 \pm 0.013$$

Do not ignore this kind of significant modification of the uncertainty value; it is necessary whenever any arithmetic process is carried out on the basic measurements.

The end result of this experiment will be a set of ℓ and T values, complete with uncertainties, and we shall then be ready to start drawing our graph.

Statistical Uncertainty

If repetition of our measuring process has shown random fluctuation in one or more of the variables, we may have decided, as described in Section 5-3, to take a sample of readings, the number of readings being chosen on the basis of the apparent magnitude of the scatter to give the precision we require. Because we must reduce the resulting set of readings to a form suitable for plotting, we must express the sample in the form: central value ± uncertainty. As described in Section 3–10, the most suitable form to choose is usually the sample mean and the standard deviation of the mean because of the readily identifiable significance of these quantities. Provided we make it clear in our report that we are quoting sample means and standard deviations of means, everyone will understand that we are specifying intervals that have a 68% chance of containing the universe mean.

While we are making these claims about the numerical significance of our measurements, we must remember the warnings given in Section 3–5. The measurement samples encountered in the work of the physics laboratory are frequently too small to permit any assessment of the actual frequency distribution of the universe from which the measurements were taken. We are therefore making an assumption when we ascribe the numerical properties of the Gaussian distribution to our sample. It is usually a good enough assumption, but we should remember that it is an only assumption.

At this point, remember also the warnings about σ estimates from small samples that were given in Section 3–11 and check that the computations are significant. In general, it is not worth using a statistical approach with fewer than 10 observations; for some particular purposes many more may be required.

It is useful to think in advance about the interpretation of the uncertainty regions on the graph. If both variables in the experiment have similar statistical character, the mean and standard deviation of the mean for each point will enable us to draw, for each point on the graph, a little rectangle whose interpretation will be clear. We may have a little more of a problem if the experiment has yielded variables of two different kinds. It is quite conceivable in, say, the experiment on free fall under gravity, which was used as an example in Section 4–2, that one variable, the distance of fall, will have an estimated uncertainty and the other will require a statistical treatment yielding standard deviations of the mean. If we were to plot values derived from these two different types of treatment, the uncertainty ranges along the two axes would be different. The uncertainty interval in one dimension would give almost 100% probability of containing the desired value, whereas the probability in the other dimension would be only 68%. It would be difficult to know how to interpret the graph, and it would be better to bring the two variables into better correspondence. Remembering that a range of twice the standard deviation of the mean gives us a 95% chance of including the universe value, we can use $2S_m$ as our uncertainty for the statistically treated variable, thus giving a range of uncertainty for each point on the graph with roughly the same significance in both dimensions.

At this stage, by one process or another, the measurement of every quantity in the experiment will have been reduced to a central value and its uncertainty, but we are not yet quite ready to start drawing the actual graph. If the graph is to be drawn with one variable on one axis and the other variable on the second axis (like load vs. extension for a spring or current vs. potential difference for a resistor), then we can proceed directly. If, however, the process of rectifying the equation for the model has led to a choice of more complicated variables for plotting (such as T^2 vs. ℓ for the simple pendulum, or h^2 vs. T^2h for the compound pendulum, etc.), we must construct these variables by some process of arithmetic computation. We obviously have no problem in performing such simple arithmetic calculations, but we must not forget that the *uncertainty* values also must be recalculated. If we are going to plot values of T^2 on the graph, the uncertainty bars or rectangles must give the actual interval over which T^2 itself is uncertain. All such computed quantities must be provided with their own uncertainty intervals, and only then are we ready to start drawing the graph.

6–3 GRAPHS

Whether the graph is to be merely an illustration of the behavior of a physical system or whether it is to be the key to assessing the experiment and calculating the answer, the aim is to set out the results in such a way that their characteristics are displayed as clearly as possible. This will involve appropriate choices of scale, proportions, and so on. First, ensure that the graph paper is large enough. It is a waste of time to plot observations having a precision of 0.1% on a piece of graph paper 18 cm × 25 cm, where a typical plotting uncertainty is perhaps 2%. As we shall see later, valuable information will be lost unless the uncertainties on the points are clearly visible, and so it is necessary to make sure that the graph paper is big enough. Second, make the graph fill the available area. This can be done by choosing the scales so that the general course of the graph runs diagonally across the paper and by suppressing the zero if necessary. When plotting the resistance of a copper wire as a function of temperature and the values run from 57 to 62 ohms, start the resistance scale at 55 ohms and run it to 65. If the scale is started at zero, the graph will look like a flat roof over a sheet of empty graph paper and convey little information.

There are times when it may be important to preserve the origin as part of the graph. It may be desirable or even necessary to examine the behavior of the graph at the zero of one or both axes. At other times, for purposes of illustration, it may be useful to show clearly the scale of some variation in relation to its zero value. However, for the purposes of the graphical analysis with which we are here concerned, it is generally best to make the graph fill the graph paper.

The method of marking each measurement on the graph paper depends to some extent on preference. One essential feature is to make sure that the range of uncertainty is clearly indicated. Only if this is done can the process of comparing the behavior of the system and the model have any meaning and the uncertainty of any future calculations of slopes, be assessed. To each point on the graph we can attach a cross with horizontal and vertical bars to indicate the range of uncertainty, or we can make each point a little rectangle surrounding the measured value and indicating by its horizontal and vertical dimensions the range of uncertainty in each coordinate. So long as the ranges of uncertainty are clearly indicated, it may not matter which method we choose; the important thing is to acquire the habit of marking uncertainties on every graph. It is also important to note on the graph itself, or in its caption, the nature of the uncertainties, whether estimated outer limits of uncertainty, statistical uncertainties of $1S$ or $2S$, or other such information. It can be very frustrating when trying to judge the significance of a graph if we have to search through the text to find out what the uncertainty marks mean. If several graphs are to be plotted on one piece of paper, make sure they are clearly distinguished by the use of some different symbol or by color or by some other means.

6–4 COMPARISON BETWEEN EXISTING MODELS AND SYSTEMS

Once all our observations have been plotted on the graph paper, we are ready to proceed with the next stage—the comparison between the properties of the system now displayed before us and the properties of any models we have available. The procedure depends on circumstances, and we describe the various situations in turn. In all the following we assume that, on account of the difficulty in representing nonlinear properties of models on hand-drawn graphs, we have chosen or rearranged our variables so that the graphs take on linear form.

Let us suppose, first, that we have a model that is fully specified and that has no undetermined quantities. The purpose of the investigation would then be only to see how well the properties of the model match the properties of the system. To do this we would simply draw on the graph, using the same scale, the graph of the function that represents the properties of the model. A typical case was illustrated in Figure 4–10, in which observations of the time of fall of an object as a function of distance are compared with the behavior of the analytical expression

$$t = 0.4515x^{1/2}$$

which represents the theoretical model of the situation.

But how are we to judge the degree of correspondence? This is where the presence of the uncertainty intervals becomes of dominating importance. If we simply plot points without uncertainty bars, the inevitable scatter in the points would mean that the probability of the line that represents the model's properties actually passing through even one (not to say more than one) of the points would be vanishingly small. So how can we say anything sensible about the outcome of the comparison? If, however, the points on the graph represent *intervals* of possibility for the location of the plotted values, it becomes possible to make logically satisfactory statements. If, as was the case in Figure 4–10, the line representing the model passed through the region of uncertainty of each point, we could say just that. Notice again that this does not mean that we have "proved" that the equation is "true," or "correct," or whatever. All we can say is that the model and the system are "consistent," or "in agreement," or "compatible," or some such phrase. As we have said before, we must make sure that we use the right language, for otherwise we may misrepresent the situation and have a good chance of misleading people. Notice also that we must be careful to say that we have found "correspondence," "consistency," "agreement," or whatever between the model and the system only at the level of precision of our experiment. Nothing in our process entitles us to ignore the fact that, at a higher level of precision of measurement, discrepancies might appear that were undetectable at the level of precision in our experiment.

Now that we have considered the case in which a model and a system turned out to have properties that are undistinguishable at the level of precision involved, we must consider the other possibilities in which the properties of the model and the system do not overlap completely.

No Detectable Discrepancy

This is the case we have already considered in detail. It is illustrated in Figure 6–1(a).

Correspondence over Part of the Range

Sometimes a model provides a satisfactory description of a system, provided the value of some variable does not exceed or fall below some limit. In this case the graphical comparison would appear as in Figure 6–1(b) or 6–1(c). An example of case (b) would be the flow of a fluid through a pipe, in which the proportionality between flow rate and pressure head is satisfactory only below the onset of turbulence. Figure 6–1(c) could be a representation of the variation with temperature of the resistivity of a metal, for which the linear model breaks down at low temperatures.

In any case that comes within this category, we would state the result of the comparison using some phrasing such as: We observed agreement (compatibility, consistency, etc.) between the model and the observations only over the range so-and-so. Or: The properties of the model and the system are observed to diverge significantly after the value such and such. Notice again that we must resist the temptation to think that something is "wrong" because we do not encounter complete correspondence between models and systems over the whole range. Both models and systems exist in their own right, and we cannot prejudge the extent to which their properties overlap. In fact, the detection of the limits on the validity of a particular model can furnish important clues for its improvement.

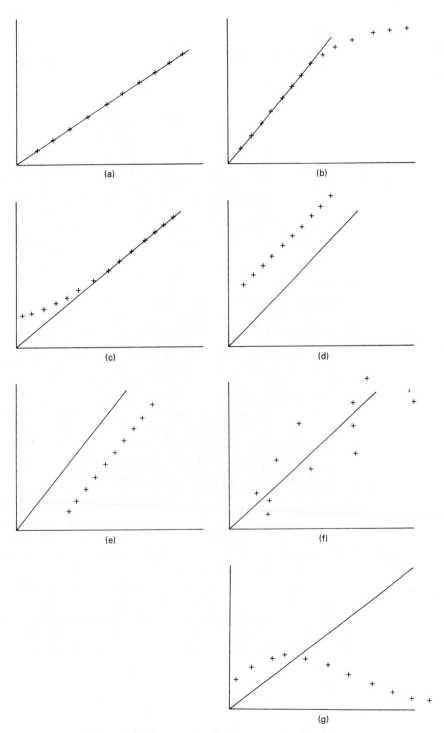

Figure 6–1 The comparison between systems and models.

Intercepts

A frequently encountered circumstance involves intercepts. The graph of the model's behavior may pass through the origin, but the observed behavior of the system may not, as illustrated in Figures 6–1(d) and 6–1(e). Such a discrepancy can arise from many different types of mismatch between the model and the system, and information about such intercepts can be helpful in analyzing experimental situations. It is usually advisable, when drawing graphs, to check the behavior of the model and system at the origin; reference was made to this point in Section 6–3 on graph drawing. As we saw in Section 4–5, the graphical analysis of an experiment is invaluable for enabling us to obtain answers that are free from the systematic errors associated with unexpected intercepts. Even with this protection, however, it is usually best to know whether an unexpected intercept exists so that we can check the overall degree of correspondence between the model and the system.

Unexpected Scatter of Points

As was described in the chapter on experiment planning, we should have carefully judged the uncertainty of our measurements before starting the experiment, and in the light of our target value for final precision, we should have made appropriate choices for our measurement methods. If we have done this satisfactorily, we shall find, on plotting the graph, that there is consistency between the scatter of the points and the uncertainties of the measurements, as illustrated in Figure 6–1(a). However, things do not always work out as we wish, and we not uncommonly find ourselves in the situation illustrated in Figure 6–1(f). It results, simply, from the presence of factors in our measurement methods that we failed to identify as we made the initial assessment of the uncertainty of the measurements.

We should not be content to leave the situation like this. It is worth checking the apparatus in an attempt to discover the cause of the fluctuation. It can be something as simple as a loose electrical connection or failure to stir a heating bath, and it is always satisfying to see such a discrepancy disappear. If for any reason it is not possible to keep the experiment going and take steps to reduce the scatter, it may be necessary to work with the results as they are and make the best statement we can about the degree of correspondence between the model and the system. We might be able to say something like: The observations are distributed uniformly about the line representing the model. For cases in which we have to obtain numerical information from lines drawn using such experimental measurements, see Section 6–7.

No Correspondence Between System and Model

We rarely encounter circumstances in which the behavior of the system bears no resemblance at all to the behavior of the model [Figure 6–1(g)]. If everything in the experiment is working as it should, this is a most unlikely outcome. Models may be in principle inadequate representations of the behavior of the physical world, but they would not be models if they were as bad as we are suggesting. Such complete failure of correspondence points clearly to an actual error in the experiment. It can be an error of interpretation of the variables, a mistake in the rectification of the equation, an error in setting up the apparatus, or a mistake in making the observations, in calculating the results, or in plotting the graph. If possible, go back to the beginning, check everything, and start again. If it is not possible to check the instruments used in the experiment, check for errors in all the analytical and arithmetic processes. If every attempt to discover an error fails, state the outcome of the experiment honestly and objectively. There is always the chance that we have discovered something new. In any case, if we are truly baffled by some failure of correspondence between a well-checked piece of apparatus and a reliable model, an honest statement of the situation is bound to be of interest to other people.

In all the foregoing we have stressed one important point: We must not think that an experiment is giving us a "right" or a "wrong" result. We just carry out the experimental process as carefully as possible and then state the outcome as honestly and objectively as we can. It is not a bad thing to be reminded from time to time that models may provide only partially satisfactory representation of the behavior of systems. It is most important for us to know the limits for the validity of models, and the manner in which models fail can furnish invaluable evidence to those who seek to improve them.

6–5 CALCULATION OF VALUES FROM STRAIGHT-LINE ANALYSIS

In all the preceding sections, we have dealt with models that were completely specified, including the numerical values of all quantities. The purpose of experiments was simply to compare the behavior of the system and of the model. As was considered in Section 4–5, however, it is also possible, indeed very common, to use straight-line analysis in determining for some quantity in the model the numerical value that is appropriate for our system. In such cases the model is not fully specified, because it contains a quantity or quantities of initially unknown value. It is not possible, therefore, to draw a graph for the model to compare with the points, for the graph initially contains nothing but the points alone, as shown in Figure 6–2(a).

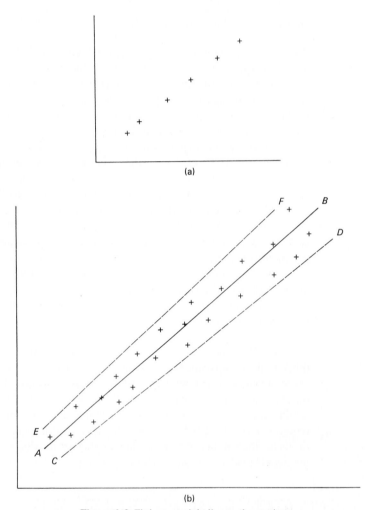

Figure 6–2 Fitting a straight line to observations.

Suppose that we have measured the values of current through, and the potential difference across, a resistor and we wish to test the observations against the model

$$V = IR$$

In the absence of a specified value for R, the behavior of the model is represented by *all* the lines on the I–V plane that have the equation

$$V = \text{constant} \times I$$

where the constant can take all values from zero to infinity. In principle, we can simply draw all these lines on the same graph as the measurements, and determine first the extent to which the behavior of the system and of the model overlap. Second, from the bundle of lines that fall within the regions of uncertainty of all the points, we can determine the range of R values that are appropriate for our system (as illustrated in Figure 4–11). For our present purpose, it is not quite as simple as that because, on the basis of the measured values shown in Figure 6–2(a), we have no right to prejudge the behavior of the system at the origin. It is best to leave the question of intercepts to a later stage and simply decide on the range of straight lines that is consistent with the observations.

There are several ways of doing this. The most satisfactory, a statistical method, is described later. In the meantime we content ourselves with simpler, mechanical procedures and carry out the time-honored practice of drawing the best straight line through the points. To do this by eye requires some mechanical aid that does not obscure half the points. An opaque ruler cannot be used, but a transparent straight edge is acceptable. Probably the most satisfactory aid is a length of dark thread, which can be stretched over the points and easily moved until the most satisfactory position is found. If difficulty is encountered in judging visually the trend of a set of points, it is often helpful to hold the graph paper at eye level and sight along the points. This makes the clustering of the points around a straight line or a systematic deviation from a straight line much clearer than in the direct view.

We can profitably identify several straight lines. The "best" straight line (whatever we mean by "best ") is one obvious candidate. In addition we can make a guess at how far we can twist that "best" line in either direction until it can no longer be regarded as an acceptable fit to the points. These two extreme positions supply us with a value for the uncertainty in the slope. If, on account of wide scatter in the points we find it difficult to identify a "best" line and its uncertainty limits, it is sometimes helpful to remember that the points and their uncertainties are really a sample from a whole band of values on the plane. The occupation of this band by the measurements may be spotty on account of the limited number of observations, and this can make it difficult to choose the lines. If this is the case, it is often helpful to imagine the band to be populated by the million or so readings that we could have made with the apparatus. We can then try to guess from the graph where the center and the edges of that band might be, and that will enable us to make our choice of lines. In Figure 6–2(b) we could have chosen AB as our "best" line, and we could have decided that the lines CD and EF would contain almost all the infinite universe of points. The lines CF and ED (not drawn) would then represent, respectively, the steepest and least steep slope of the set of lines that are consistent with the observational points.

Once we have chosen the lines, we can set about determining the numerical value of their slopes so that we can obtain the answer we want, such as, in the case of the $V = IR$ example, the value of R. For our purpose, the question of slope has nothing to do with the angle made by the lines on the graph paper; we are talking about intervals of the measured variables I and V, and so the slopes must be calculated analytically. For a line such as AB on Figure 6–3 look carefully near the ends, and identify as exactly as possible places at the top and bottom of the line where it crosses an intersection of lines on the graph paper. Identify the coordinates (I_1, V_1) and (I_2, V_2) of these intersections and evaluate the slope as

$$\text{slope} = \frac{V_2 - V_1}{I_2 - I_1}$$

We then immediately have

$$R = \text{slope}$$

which gives us the answer we want. In more complicated expressions, the value for the slope may give the desired answer only after computation with other measured quantities.

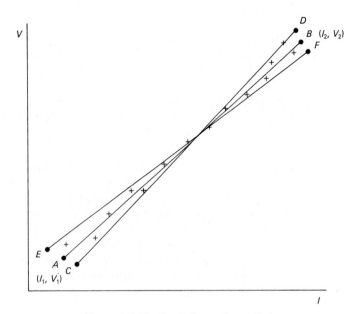

Figure 6–3 The "best" slope and outer limits.

We carry out this process three times. The line AB gives us our chosen "best" value for R, and the other two lines, CF and ED, will give us the upper and lower limits, outside which we are "almost certain" the value for R does not lie. It usually happens that these extreme values for the uncertainty range are roughly equidistant from the central value, and is then finally possible to state the value for R as

$$R = \text{value} \pm \text{uncertainty}$$

It may sometimes appear that the "best" line and the two outer limiting lines are not equally spaced. The reason is usually that the graph contains too few points to allow good assessment of the positions of the lines. Although circumstances occasionally appear in which experimenters feel obliged to express a result as

$$\text{value}\begin{Bmatrix} +\text{uncertainty 1} \\ -\text{uncertainty 2} \end{Bmatrix}$$

visual judgment of a graph is rarely sufficiently precise to justify such a procedure. If it is genuinely difficult to identify a "best" line, it is acceptable to delineate the edges of the band of values (lines ED and CF in Figure 6–3) and to calculate simply the maximum slope (of the line CD) and the minimum slope (of the line EF). We can then give the experimental answer as the interval between these two slopes, or else we can calculate a central value for the slope as the average between these two extreme values with an uncertainty equal to $\pm$ half that interval.

If in our experiment the desired answer is not equal to the slope directly, the expression for the slope may contain a number of quantities, and the unknown may have to be calculated from the slope by a separate arithmetic process. If these other quantities are themselves uncertain, the uncertainty in the answer will have to be found by combining the uncertainty of the slope with the other uncertainties, using the techniques of Chapter 2.

It is natural at this stage to think about the significance of the uncertainty associated with quantities obtained from graphs. The significance depends on the type of uncertainty marked on the graph. If the bars indicate outer limits of possible variation (subjectively assessed or $2S_m$ in the case of statistical fluctuation), then the limits on the slope will similarly be of this nature. If the points have been marked with $1S_m$ limits, the limiting slopes CF and ED will probably represent limits implying better than 68% probability, because the limiting lines are drawn with a pessimistic bias.

We have assumed in the foregoing that the scatter encountered in the actual results is within the predicted range of uncertainty. If this is a valid assumption, the use of the limiting lines gives rise to a fairly well-defined value for the uncertainty in slope. If, however, the scatter is well outside the expected range of uncertainty (owing to an unsuspected source of fluctuation), then there may be no unique setting for lines within which we are "almost certain" the answer lies. In such a case and in all precise work, there is no substitute for the method of least squares, which is described in Section 6–7.

In choosing our three lines, we have deliberately excluded the origin as a factor in making the choice, precisely because the behavior of the system at the origin may be one of the things we wish to examine. If the graph of the model's behavior does pass through the origin, we should inspect the three lines in that region. It is most unlikely that our central line will pass exactly through the origin, but if the area between the two limiting lines does include the origin, we can say that we have consistency between the model and the system, at least at the present level of precision. Only if both limiting lines clearly intersect an axis on one side of the origin can we claim that we have unambiguously identified an unexpected intercept.

If the behavior of the model does lead us to expect an intercept from which we hope to obtain the value of some quantity, the intersection of the three lines on the axis in question will give us that intercept directly in the desired form: value ± uncertainty.

6–6 CASES OF IMPERFECT CORRESPONDENCE BETWEEN SYSTEM AND MODEL

When the correspondence between model and system is only partial, we must be careful to obtain answers without introducing systematic error from the discrepancies. Refer to Figures 6–1(b) and 6–1(c) and consider first the cases in which the measured values correspond adequately with the straight line of the model over only a limited range. Obviously, our evaluation of slopes should be confined to those regions in which the system and model are compatible. The points that deviate systematically from the line clearly arise from physical circumstances that are not included in the model, and it is obviously inappropriate to include them in any calculations that are based on the model. We disregard, therefore, all points that deviate systematically from straight-line behavior by an amount clearly in excess of the estimated uncertainties and observed scatter of the points, and we restrict our calculations of the slope and its uncertainty to the linear region.

A second point concerns intercepts. Even if the model's behavior passes through the origin, it is not uncommon to find that the graph shows an intercept. Such a deviation can arise from a variety of causes; fortunately, many of these prove to be harmless. If the discrepancy causing the intercept affects all the readings in the same way (like an undetected zero error in an instrument or a spurious and constant emf in an electrical circuit), then the graph will give a slope that is free of the systematic error that would otherwise be introduced. It is wise, therefore, to arrange the experiment so that the answer will be obtainable from the slope of the graph, whereas quantities that may be subject to undetermined systematic error should be relegated to the role of intercepts. The capacity of graphical analysis to provide answers that are free from many types of systematic error is one of its chief advantages.

6–7 THE PRINCIPLE OF LEAST SQUARES

All the procedures described in the preceding sections have one characteristic in common; they are all based on the use of visual judgment by the experimenter. Thus, although the procedures are commonly used and are useful, they are vulnerable to the criticism that, even when they are carefully carried out, we cannot be sure of the numerical significance of the results. It would be comforting if we could use some mathematical procedure to identify the "best" line for a set of points, for then we would be released from the insecurity of personal judgment. In addition, we could hope to find out what we mean by "best" and to assess the precision of that choice.

The procedure in question is based on the statistical principle of **least squares**. We discuss the procedure mostly in the restricted application to straight-line fitting to measured values. It is possible, in addition, to use the least-squares principle to fit other functions to sets of observations, and this is briefly considered later. For the moment, however, we restrict our attention to straight lines only, so that the discussion gives a clear and simple illustration of the principle. Further detail on the principle in general is in Appendix 2.

Consider that we have a set of N values of a variable y measured as a function of a variable x. We must restrict ourselves to the special case in which all the uncertainty is confined to the y dimension; that is, we shall assume that the x values are known exactly or at least so much more precisely than the y values that the uncertainty in the x dimension can be neglected. Fortunately, many common experimental situations involve one variable which, if not exactly zero in uncertainty, is at least so much more precise than the other that the assumption we are making is good enough. If this condition cannot be satisfied, the simple treatment following is not valid. The least-squares method can be extended to cover the case of uncertainty in both dimensions, but the procedure is not simple. Anyone who wishes to pursue the subject can find an excellent treatment in the text by Wilson listed in the Bibliography.

The questions now to be answered by our mathematical procedure are: Which of all possible lines on the x–y plane do we choose as the best line? What do we mean by "best"? The principle of least squares makes this choice on the basis of the deviations of the points in a vertical direction from a line. Let AB in Figure 6–4 be one candidate for the status of "best" line. Consider all the vertical intervals between the points and the line, of which P_1Q_1 and P_2Q_2 are typical. We define the best line to be the one that makes the sum of the squares of deviations such as P_1Q_1 and P_2Q_2 a minimum.

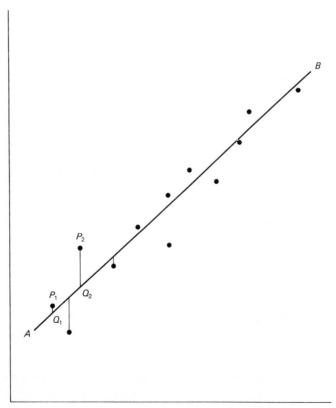

Figure 6–4 Fitting a straight line to a set of points by the principle of least squares

Notice that we have no right to consider an invented criterion like this as providing any automatic path to "truth" or to "correct" answers. It is simply one choice of a criterion for optimizing the path of the line through the points. It does, however, offer some advantages over other possibilities, such as minimizing the third power of the intervals, or the first power, and so on. Although we need not in general be concerned with the rationale for the principle of least squares as we use it, the basis for its claim to validity is of interest. It can be proved that the procedure of minimizing the *squares* of the deviations gives rise on repeated sampling to smaller variance of the resulting parameters, such as slope, than does the use of any other criterion. We consequently have greater confidence in results obtained by using the principle of least squares than is the case for any competitor. As a result, use of the principle of least squares is almost universal.

We now put the least-squares principle into mathematical form. We define the best line to be that which leads to the minimum value of the sum

$$\sum (P_i Q_i)^2$$

and we desire the values of the parameters, slope m and intercept b, of that best line.

Let the equation of the best line be

$$y = mx + b$$

The magnitude of the deviation $P_i Q_i$ is the interval between a certain measured value y_i and the y value of the point, on the line, at that value of x. This y value, Y_i, can be calculated from the corresponding x value by using

$$Y_i = mx_i + b$$

so that, if we call each difference δy_i, we have

$$\delta y_i = y_i - Y_i$$
$$= y_i - (mx_i + b) \tag{6-1}$$

The criterion of least squares then enables us to obtain the desired values of m and b from the condition

$$\sum [y_i - (mx_i + b)]^2 = \text{minimum}$$

Write

$$\sum [y_i - (mx_i + b)]^2 = M$$

Then the condition for the minimum is

$$\frac{\partial M}{\partial m} = 0 \qquad \text{and} \qquad \frac{\partial M}{\partial b} = 0 \qquad (6\text{-}2)$$

A brief algebraic exercise (given in full in Appendix A2–2) then allows us to obtain the values of slope and intercept for the best line as

$$m = \frac{N \sum (x_i y_i) - \sum x_i \sum y_i}{N \sum x_i^2 - \left(\sum x_i\right)^2} \qquad (6\text{-}3)$$

and

$$b = \frac{\sum x_i^2 \sum y_i - \sum x_i \sum (x_i y_i)}{N \sum x_i^2 - \left(\sum x_i\right)^2} \qquad (6\text{-}4)$$

We have now succeeded in replacing the sometimes questionable use of personal judgment by a mathematical procedure that leads to results of well-identified significance and universal acceptability. In addition, because there is some statistical meaning in the new method, we can expect a more precise form of uncertainty calculation. The least-squares principle allows us immediately to obtain values for the standard deviations of the slope and the intercept, giving us uncertainties of known statistical significance.

The standard deviations of the slope and intercept are calculated in terms of the standard deviation of the distribution of δy values about the best line, which we call S_y. It is given by

$$S_y = \sqrt{\frac{\sum (\delta y_i)^2}{N - 2}} \qquad (6\text{-}5)$$

(Do not worry about a standard deviation being calculated with $N - 2$ in the denominator instead of the familiar N or $N - 1$; it is a consequence of applying the definition of the standard deviation to the positioning of a line on a plane.) The values for the standard deviation of the slope, S_m, and for the standard deviation of the intercept S_b can then be calculated to be

$$S_m = S_y \times \sqrt{\frac{N}{N \sum x_i^2 - \left(\sum x_i\right)^2}} \qquad (6\text{-}6)$$

and

$$S_b = S_y \times \sqrt{\frac{\sum x_i^2}{N \sum x_i^2 - \left(\sum x_i\right)^2}} \qquad (6\text{--}7)$$

The full derivation of these equations can be found in Appendix 2.

These values for the standard deviations can be used in association with the values of m and b to determine intervals with the normal meaning, namely, that intervals of one standard deviation give a 68% chance of enclosing the universe value, two standard deviations 95%, and so on. One important advantage of the least-squares method is, therefore, that it supplies statistically significant values for the uncertainties in our slope and intercept. Not only are they statistically significant; they are in addition derived objectively from the actual scatter in the points themselves, irrespective of any optimistic claims for the uncertainties of the measured values.

Besides the complete mathematical description of the least-squares method in Appendix 2, there is also an extension to the method that we have excluded from the present discussion. If, in the experiment, the points used in the least-squares calculation are not equally precise, we should use some procedure that accords greater importance to the more precise measurements. This procedure is called **weighting**. The use of weighting is not limited to straight-line fitting. The procedures are applicable whenever we wish to combine observations in some way, even in such a simple process as determining the mean of a set of values of unequal precision. The equations for finding a weighted mean and for doing a weighted least-squares calculation are given in Appendix 2.

6–8 LEAST-SQUARES FIT TO NONLINEAR FUNCTIONS

The procedures used in Section 6–7 to determine the slope and intercept of the best straight line can, in principle at least, also be applied to nonlinear functions. We can write an equation analogous to Equation (6–1) for any function, and we can still use a requirement similar to Equation (6–2) to express the minimizing of the quantity M with respect to the parameters in our chosen model. If the resulting equations for the parameters are easy to solve, we can obtain their values just as we did for straight lines.

Frequently, however, it is not easy to solve the equations. In such cases we abandon the attempt to obtain an analytical solution to the problem and rely on the computer to provide us with approximate solutions by using iterative techniques. We construct a trial function, calculate the sum of the squared differences, and then vary the chosen function until a minimum is found for that sum. Descriptions of such computer-based methods are given in the text by Draper and Smith that is listed in the Bibliography. If, however, a method can be found to test a model in linear form, this is certainly simpler.

In all cases, it is the responsibility of the experimenter to choose the type of function to be used. All that the least-squares method can do is to give us, for whatever function we choose, those values of the parameters that provide the best fit with the observations.

6–9 PRECAUTIONS WITH LEAST-SQUARES FITTING

The mathematical procedures for least-squares fitting are completely objective and impartial. The use of Equations (6–3) and (6–4) for linear fitting drives a straight line through *any* set of points with complete disregard for the appropriateness of a straight-line function. If, for example, an experiment has given us a set of observations (Figure 6–5) that clearly show the breakdown of a linear model and we heedlessly use the least-squares procedure on the whole set of observations, we shall obtain the parameters of a line, *AB*, that has no significance at all, neither for the model nor the system. Unthinking use of the least-squares procedures must be studiously avoided.

This warning is all the more important because of easy access to calculators and computers that can give, at the touch of a few buttons, least-squares parameters for any set of numbers we care to insert. We *must* remember that, if we are comparing straight lines with our set of observations, it is because *we* have made the decision that this is a reasonable thing to do. We must not, therefore, even contemplate using a least-squares procedure until we have plotted the observations on a graph and satisfied ourselves, by visual inspection and personal judgment, that linear fitting is appropriate. In addition, as has been mentioned, it may be necessary to decide that some of the observations fall outside the scope of the model and are not appropriate for inclusion in the choice of the best straight line. Only after we have carefully considered the whole situation graphically and visually and are sure that linear fitting is appropriate over all or part of the range of the observations, are we justified in starting the least-squares procedure. Failure to heed this warning can give rise to serious error in experiment interpretation.

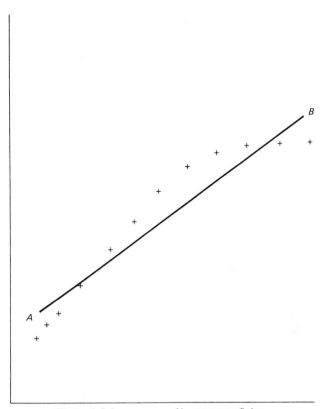

Figure 6–5 Improper use of least-squares fitting.

6–10 FUNCTION FINDING

All the foregoing discussion involved the assumption that we were already in possession of a model that we wished to compare with a system. Although this is commonly the case, it does happen sometimes that we have a set of observations for which no model is available—for example, in research on a phenomenon that has never been observed before, or in work on a system that is so complex that a theoretical model will never be available. The observations, when plotted in elementary form, will probably show a curve of no readily identifiable form. In the absence of a model, what are we to do?

One thing we can do is try to find functions that have some degree of correspondence with the observations. Such a procedure can be very useful. For example, in very complex systems for which there is little hope of constructing theoretical models, it may be the only thing we can do. An empirical model, even if is only a mathematical function that is nothing more than a restatement in mathematical form of the actual behavior of the system, can facilitate computer processing of the observations and is indispensable for procedures such as interpolation or extrapolation. Such models can be used, for example, to predict the response of a country's GNP to a change in taxation or to obtain measurements of temperature from the calibration curve of a resistance thermometer.

In simpler systems for which some hope of constructing a theoretical model from first principles exists, some functions, if shown to be appropriate to the observations, may be able to give valuable guidance in model building by suggesting the type of physical process involved in the phenomenon. Even so we must be careful. The fact that we have identified a function that seems to be consistent with our set of observations at a particular level of precision does not "prove" that we have found the "right" function. Quite often, functions of widely varying type can show closely similar variation, especially over a short range of the variables, and "guidance" from an inappropriately identified function can be misleading. It can retard genuine theoretical progress for years. The history of physics contains many examples of such failure to understand that any choice of an empirical function must be provisional.

With due attention, therefore, to the possibly limited significance of our procedures we describe some of the methods used. They can be quite simple in a few cases, and two of these are important because they involve functions that are of relatively common occurrence. Assume that we have made measurements of two variables that we can call x and y.

Power Law

In the discussion of experiment planning, we have already described the nature of logarithmic plotting and the uses to which it can be put. For the sake of completeness in the topic of experiment evaluation, we briefly recapitulate that description in this and the following section. Consider the function

$$y = x^a$$

where a is a constant. We have

$$\log y = a \log x$$

(where the logarithms can be taken to any base we please) and a graph of $\log y$ versus $\log x$ is a straight line with slope a. Consequently, if we wish to test whether a power law is a good function for our observations, we can plot them in the form log y versus log x. If the resulting points plotted in this way correspond well with a straight line, we can say that a function involving a simple power, positive or negative, integral or fractional as determined by the graph, is a good fit with our observations. The value of the appropriate power, a, is derived from the slope of the graph and is obtained within uncertainty limits that depend on the uncertainties plotted on the points. A graph like this can be plotted on ordinary graph paper by plotting the actual values of log x and log y, or we can use logarithmic graph paper. This paper has rulings that are spaced in proportion to the logarithms of the numbers, so that we can plot our observations directly on the paper.

Exponential Functions

For many physical phenomena an exponential function is appropriate. Consider

$$y = ae^{bx}$$

where a and b are constants. In this case

$$\log_e y = \log_e a + bx$$

(where the logarithms *must* be natural logarithms, taken to the base e), and the graph of the function is a straight line when we plot $\log_e y$ versus x. If there is reason to suspect that an exponential function is appropriate to a particular system, we should do a semi-log plot, either on ordinary graph paper by obtaining the values of $\log_e y$, or on semi-log graph paper, which has one logarithmic and one linear scale. The appropriate values of a and b are obtainable from the intercept and slope of the line, with uncertainties determined by the plotted uncertainties of the measured values.

6–11 POLYNOMIAL REPRESENTATION

If neither a simple power nor an exponential function has been found to provide a good match to a set of observations, the probability of stumbling on a more complicated function that would be appropriate is very small. In such cases it is often useful to resort to a polynomial representation, which we can write in the form

$$y = a_0 + a_1 x + a_2 x^2 + \ldots$$

A recourse of this nature is really an admission that we have no idea what is going on in the system. Although such a representation may not contribute much insight regarding the fundamental theoretical basis for the operation of the system, it at least offers some of the advantages of empirical models. If nothing more, it allows computerized processing of our knowledge of the system's behavior and provides a satisfactory basis for interpolation and extrapolation.

The coefficients in such a polynomial expansion that make it appropriate for our particular system can be found by using the least-squares principle. Recalling the remarks of Section 6–8, it will be appreciated that the associated difficulties escalate rapidly with the number of terms that are required in the polynomial to give satisfactory correspondence with the observations. A fuller discussion of such methods is in the text by Draper and Smith that is listed in the Bibliography.

A similar method is available if the scatter in the observations is not too severe and if the highest precision is not required. The techniques of the calculus of finite differences can be applied to the observations, and a difference table can be used for interpolation and extrapolation or for polynomial fitting. A complete discussion of difference-table methods is in the texts by Whittaker and Robinson and by Hornbeck that are listed in the Bibliography, and an elementary description is in Appendix 3.

6–12 OVERALL PRECISION OF THE EXPERIMENT

At the beginning of the experiment, we guessed at the uncertainties that we were likely to encounter. This was only an estimate made for the purpose of supplying guidance for the conduct of the experiment. At the end of the experiment, we should look back and, by critical assessment of the results, evaluate the precision actually achieved. It does not matter very much what type of uncertainty we choose—estimated range of possible value, standard deviation, standard deviation of the mean, and so on—provided only that we clearly state the kind of uncertainty we are quoting.

To be useful, the overall uncertainty figure must be realistic and honest, even if the outcome of the experiment is less favorable than we had hoped. It should also include all identifiable sources of uncertainty. If the balance point cannot be identified within 2 to 3 mm or if errors are introduced by nonuniformities of the slide wire, there is no point in claiming that potentials read on a 1 m slide-wire potentiometer are precise to 0.2% simply because the scale is graduated in millimeters.

Known contributions from systematic errors should not be included at this stage, because the appropriate corrections to the measurements should already have been made. On the other hand, a source of systematic error, whose presence we suspect but whose contribution cannot be evaluated accurately, should be described and appropriate allowance made in the overall range of uncertainty. The final statement depends on the circumstances.

Result Is the Mean of a Set of Readings

The best quantity to quote is the standard deviation of the mean, because it has recognizable numerical significance. Sometimes the standard deviation itself is quoted. It is always essential to quote the number of readings so that the reliability of the σ estimate can be judged.

Result Is the Consequence of a Single Calculation

In the undesirable event that no graphical analysis has been possible and the result is obtained algebraically from a number of measured quantities, use the methods of Chapter 3 to calculate either outer limits for the uncertainty or else a standard deviation.

Result Is Obtained Graphically

If the straight line has been established by a least-squares method, the uncertainties in the constants m and b will have been obtained directly. These uncertainties have the advantage that they have been obtained from the actual scatter of the points, regardless of their estimated uncertainties. (This does not mean that, if we intend to make a least-squares fit to a straight line, we should not bother to plot the uncertainties or even not draw a graph at all. As was emphasized in Section 6–9, the graph, with the uncertainties on the points, is still needed to judge the range of matching between the model and the system before the least-squares calculation is done.) If the straight line has been drawn by eye, the lines at the limits of possibility will give the possible range of slope and intercept. This uncertainty in slope may have to be combined with the uncertainties of some other quantities before the final uncertainty of the answer can be stated.

As mentioned earlier, it probably does not matter much what kind of uncertainty is quoted, so long as one is quoted and the nature of the quoted value is made clear. Also, when one is working through lengthy uncertainty calculations, the arithmetic may be simplified by dropping insignificant contributions to the total uncertainty. There is no point in adding a 0.01% contribution to one of 5% because we do not know the 5% with three-figure accuracy. In the final statement of uncertainty, it is not commonly valid to quote uncertainties to more than two significant figures; only work of high statistical significance justifies more.

Once the overall uncertainty of the final answer has been obtained, the question of the number of significant figures to be retained in the answer can be considered. This matter has already been covered in Section 2–11; we repeat the discussion here simply for the sake of completeness as we discuss experiment evaluation.

There is no unique answer to the question of significant figures, but in general, one should not keep figures after the first uncertain figure. For example, 5.4387 ± 0.2 should be quoted as 5.4 ± 0.2, because if the 4 is uncertain, the 387 are much more so. However, if the uncertainty is known more precisely, it might be justifiable to keep one more figure. Thus, if the uncertainty were *known* to be 0.15, it would be valid to quote the answer as 5.44 ± 0.15.

If a measurement is quoted with a percentage precision, the number of significant figures is automatically implied. For example, what could be meant if a measurement were quoted as $527.64182 \pm 1\%$? The 1% means that the absolute uncertainty could be calculated to be 5.2764. The precision itself, however, is quoted to only one significant figure (1%, not 1.000%), so that we are not justified in using more than one significant figure in the absolute uncertainty. We shall call the absolute uncertainty 5, and this implies that, if the 7 in the original number is uncertain by 5, the 0.64182 has no meaning. The measurement could then be quoted as 528 ± 5 or, alternatively, $528 \pm 1\%$. If a set of readings has yielded a mean as the answer, the number of significant figures in the mean will be governed by the standard deviation of the mean, and the number of significant figures in the standard deviation will be governed in turn by the standard deviation of the standard deviation.

Finally, always be sure to quote an answer and its uncertainty in such a way that the two are consistent—that is, neither as 16.2485 ± 0.5 nor as 4.3 ± 0.0002.

6–13 THE CONCEPT OF CORRELATION

Until now we have been considering the interpretation of experimental results in which relatively precise observations were available and the models were relatively satisfactory. We are not always so lucky, and much of modern experimenting is less simple and clear-cut than the preceding sections might suggest. In many areas of science it is common to be concerned with subtle phenomena in which the effects we seek to measure can be wholly or partially masked by statistical fluctuation or other perturbations. In such cases, far from being able to make detailed comparisons between the system and a model, we may find it difficult to obtain clear-cut evidence that the effect we are considering even exists. This is a not uncommon situation in, for example, biological, medical, and environmental studies. We are all familiar with the discussions about the role of smoking in lung cancer, of low levels of ionizing radiation in leukemia, or of diet in cardiovascular disease. In such cases the concept of "proof" is almost always brought into the discussion in phrases such as: We have not proved that smoking causes lung cancer. Can we prove that heart attacks are less likely if we eat margarine instead of butter? And so on. In cases like these we are in a very different area of operation from our earlier kind of experimenting, and it is worthwhile spending a moment to think about what we mean by words like "proof" and "cause."

Consider two experiments. One might be a measurement of the current through a resistor as a function of potential difference across it, and the result might be as shown in Figure 6–6(a). In this experiment, have we "proved" that the current is "caused" by the potential difference? Certainly the current at the top end of the range is different from that at the low end by an amount greatly in excess of the uncertainty of measurement, and that gives us confidence that the variation actually existed. Given that it existed, was it "caused" by the change in potential difference? On that one occasion we certainly did observe that the current did increase as the potential difference increased. However, it could be that the current has nothing to do with potential difference and that the increase in current was caused by some totally separate factor, such as atmospheric pressure. The apparent relationship with potential difference could have been totally accidental. Philosophers for hundreds of years have been warning us that events observed to take place simultaneously are not necessarily causally related. In the present case, however, accumulated experience with the experiment, using multiple repetition and careful attention to the control of other variables, will gradually convince us that potential difference and current are genuinely related. Only a philosophical purist would quarrel with the claim that the potential difference caused the current to flow.

The situation is quite different in less clear-cut cases. Another experiment might yield the result shown in Figure 6–6(b). This result would be likely if we were dealing with, perhaps, the number of colds experienced by the whole student body of a university as a function of the amount of ascorbic acid ingested daily. Can we say that the number of colds is dependent on ascorbic acid dose or not? We might conduct a well-designed experiment using an experimental group of 100 students who were given ascorbic acid every morning and [as described in Section 5–6(b)], a control group of 100 students who unknowingly swallowed sugar pills instead of ascorbic acid every morning. If we find that the control group had 125 colds in a particular lar period and that the experimental group who took ascorbic acid had a total of

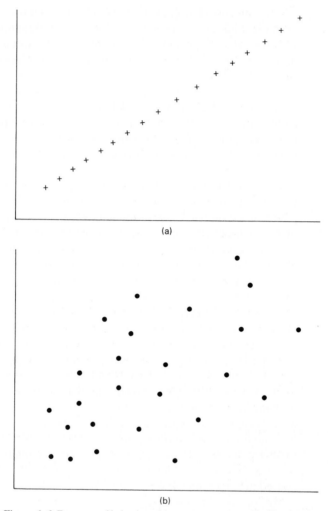

(a)

(b)

Figure 6–6 Extremes of behavior with respect to cause-and-effect relationships.

106 colds, the questions we must ask are: Is this difference significant? What do we mean by "significant"? If the difference is significant, can we attribute it to the ascorbic acid? And so on. Even painstaking attention to the details of experimenting, control over samples, elimination of extraneous variables, repetition of the experiment, and the like, may not clear up the situation very much. Biological systems are so complex that we can rarely attain the degree of control over variables that characterized the electrical experiment. It therefore becomes inappropriate to seek the kind of "proof" that is available in other systems. We cannot say that we have "proved" that smoking causes

lung cancer or that ascorbic acid reduces the incidence of colds in the same way that we can "prove" that a potential difference "causes" a current to flow. We have to be content with another class of statement, which, although less exact, can still be adequately significant and completely convincing.

This type of statement can be illustrated by reference to a diagram such as Figure 6–7. These measurements were made to test the proposition: The number of counts obtained from a weak radioactive source depends on the length of time of counting. Here, statistical fluctation is almost as big as the effect we seek to observe, but we can still see that there is an upward trend in the observations and we say that there exists a **correlation** between one variable and the other. This means that we can observe a tendency for one variable to follow the other, although fluctuations arising from other factors prevent the observation of a unique, one-to-one correspondence. The mathematical study of such correlation is called **regression analysis,** and it supplies a numerical measure of the degree of correlation between two variables that we call a **correlation coefficient.**

We encounter the concept of correlation in two significant cases: (1) if, of two measured variables, one can be regarded as the cause of the other but its effect is partially masked by random fluctuation, and (2) if two variables can be regarded as simultaneous consequences of a common cause whose effect, as before, is partially obscured by random fluctuation. In either case we may be able to say that we can observe a certain degree of correlation between one variable and the other.

The mathematical properties of correlated variables are described in the standard texts on statistics. We confine ourselves here to quoting the equation by which one calculates correlation coefficients. For a pair of measured variables x and y, the expression for the correlation coefficient, commonly called r, is

$$r = \frac{\sum xy}{\sqrt{\sum x^2 \sum y^2}}$$

Values of r calculated in this way can vary all the way from 1, for perfect, fluctuation-free dependence of one variable on another, to 0 if there is no connection whatsoever between the two variables. For intermediate values, the correlation coefficient indicates the extent to which the observed variation in one quantity can be ascribed to variation in another. In case (1) it is the extent to which the variation in the output variable can be ascribed to variation in the input variable; in case (2) it is the extent to which the vari-

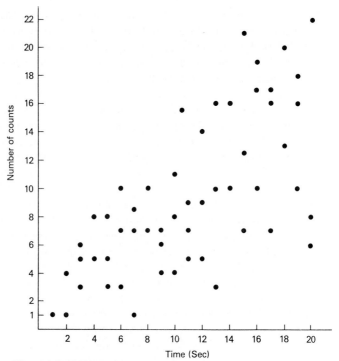

Figure 6–7 The dependence of number of counts on counting time for a weak radioactive source.

ation in both variables can be ascribed to variation in whatever is the common source of influence. In the circumstances commonly found in the type of experimenting we discussed in this text, values of r close to unity are usual.

Even when we do observe a correlation, we must still be careful about inferring *causal* connection between the various variables. If we observe that one variable seems to correlate well with another, we have not "proved" that one variable "causes" the other in the same sense as we used these words in the example about electric current. The literature has many examples of false and misleading correlations. One conference speaker illustrated this point with a tongue-in-cheek claim to have discovered the cause of cancer. He showed a graph of a quantity that correlated beautifully with the increase in some type of cancer and only later revealed that the other variable was the consumption of fuel oil in the British Navy. In another case, intended apparently to be taken seriously, a 1920s newspaper report described the "discovery" of the cause of polio, because the incidence of the disease correlated so well with the number of motor cars on the roads.

However, such amusements do not discredit the study of correlations or the search for causal relations; they merely serve as another reminder of the need for caution and clear thinking. When treated with great care, and especially when the correlation can be observed repeatedly, correlation studies can and do supply convincing evidence of causal connection. Because of the immense importance that many of such issues have in public affairs, it is important to have a clear understanding of the nature of correlation and the methods available for significance testing. Further discussion is beyond the scope of this volume, but pursuit of the topic in the texts on statistics (listed in the Bibliography) is earnestly recommended.

6-14   USE OF COMPUTERS IN EXPERIMENT EVALUATION

Many software applications have facilities that aid in processing experimental observations. Apart from word processing, which is obviously helpful in the preparation of reports (considered in the next chapter), any application that contains a spreadsheet facility eases the burden of arithmetic calculation enormously. Advanced spreadsheet programs such as Lotus 1-2-3 and Quattro Pro have built-in mathematical functions that can cope with all but the most specialized requirements. Nevertheless, some limitations and precautions must be kept in mind. We consider a number of different aspects in turn.

Graph Drawing

Almost all spreadsheet programs produce attractive graphs of experimental observations, but their use in serious experimenting is limited. As has been stressed repeatedly, the function of a graph in experimenting is to enable the experimenter to form a judgment about the degree of correspondence between the system and the model. A graph can do this only if it is large enough to show clearly both the scatter of the points and the uncertainty bars or boxes that have been plotted on them. This can require large sheets of graph paper for an experiment containing precise measurements, but spreadsheet programs and normal printers usually produce graphs on 8 1/2-by-11-inch paper. These may be acceptable as illustrations in a report, but they are rarely satisfactory for accurate analysis of experiments. Unless the computer can produce output on a large plotter, there is no substitute for a good, big, hand-drawn graph.

(1) Uncertainties.   Apart from size, a problem appears with the common spreadsheet programs when we try to represent uncertainties on the points. Both Lotus 1-2-3 and Quattro Pro have an alternative to the normal "x–y" choice on the graph menu. They call it "High-Low," and it is designed to represent stock market prices. It produces a vertical bar to represent the plotted values and so is suitable for representing experimental observations that have uncertainty in the vertical direction only. Figure 6–8 shows the observations from Table 4–1 in a graph drawn using Quattro Pro 4.

On the work sheet the load values were entered in column A. In column B were entered the values for the extension incorporating the minus value of the uncertainty, and in column C were entered the extension values calculated with the positive sign for the uncertainty. For the graph choices, the primary selection of graph type was "High-Low." Columns B and C were selected as the "1st Series," and column A was selected as the "X-Axis Series." Such a representation is satisfactory if all one wants is to show the observations and if the uncertainty is restricted to the vertical direction. The "High-Low" option for graph type, however, does not allow us to plot graphs of functions in any easy way, and we are restricted to showing experimental values only.

To plot functions easily we must use the ordinary "x–y" choice for graph type, as opposed to the "High-Low." Figure 6–9 shows the observations from Table 4–3 as re-plotted using Quattro Pro 4.

Because we wish to plot the graph of the model as well as the observations, we must use the "x–y" choice for graph type rather than the "High-

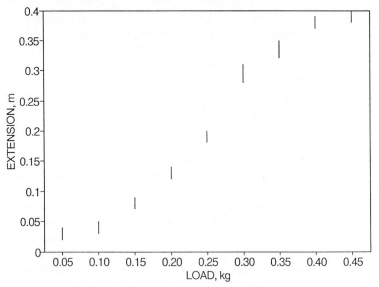

Figure 6–8 The observations of Table 4–1 re-plotted using Quattro Pro.

TIME vs. SQUARE ROOT OF DISTANCE

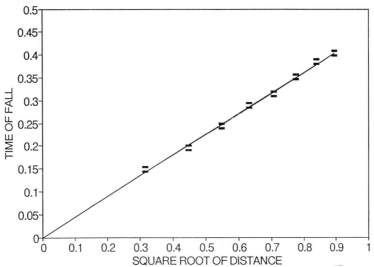

Figure 6–9 The observations in Table 4–3 re-plotted using Quattro Pro

Low" form, which raises problems about representing the uncertainty on the points. The diagram was prepared as follows. In the work sheet, column A contains the values of distance. Columns B and C contain the measured values of time of fall using, respectively, the negative and positive signs for the uncertainty. Column D contains values of the square root of each distance. To draw the graph of the experimental observations, columns B and C were used together as the "1st Series" (i.e., the y-axis variable), and column D was used as the "X-Axis Series." Under the menu entry "Customize Series," the "Format" was chosen to be "Symbols" (so that no connecting lines would be drawn between the points), and the "Marker" selected was "Horizontal Line." This produced an indication on the graph of the vertical range of each measured point. To draw the graph of the model function, column E was used to calculate values of time for each distance of fall, using the model function. That column was then chosen as the "2nd Series" under the graph menu and plotted by using "Lines" as the format. In this way both the experimental observations and the model function can be shown simultaneously on the same graph. Such a diagram, although too small to allow accurate appraisal of the experiment, can serve as an acceptable illustration in a report.

(2) Least Squares Analysis.   The larger spreadsheet programs such as Lotus 1-2-3 and Quattro Pro, have built-in facilities that use the least-squares principle for fitting straight lines to sets of observations. These facilities can make the process fast, accurate, and simple—almost too simple, in fact—and it is essential to bear in mind the warning given earlier about proceeding to least-squares analysis only after adequate personal appraisal of the situation has been made by using an appropriately large graph.

To do a least-squares analysis, we enter the observations in two columns and then choose the option marked "Regression." This immediately supplies values for the slope and intercept of the best line, as well as other information that we discuss in a moment. We usually want to illustrate also the best line on the graph of the actual observations, and we do this by calculating values that lie on the best line. In Figures 6–10(a) and 6–10(b) we show both a printout of the work sheet and the resulting graph for the observations listed in Table 4–3.

In the work sheet, Column A contains the measurements on distance. Columns B and C contain the measurements of time calculated by using the negative and positive values of the uncertainty. Column D gives the calculated values of the square root of the distance. The regression analysis was done on the basis of these four columns. In Quattro Pro the regression facility is found in the "Tools" menu under "Advanced Math." Using the "Regression" option, column D was selected as the "Independent Variable," and columns B and C together were selected as the "Dependent Variable." The output of the regression calculation is shown below the tabulated values. The meaning of the various items is as follows:

"Constant"	means	the intercept on the y-axis
"X Coefficient"	means	the slope
"Std Err of Coef"	means	the standard deviation for the slope

and these three represent a large part of what we require from the regression analysis. The one remaining item, the standard deviation for the intercept, is not represented directly in the regression output, but it can be obtained by a short calculation. The quantity "Std Err of Y Est" means the standard deviation of the y values about the best line. This is the quantity that, in Section 6–7 we called S_y, and the value we need, the standard deviation of the slope, was given by

$$S_b = S_y \times \sqrt{\frac{\sum x_i^2}{N \sum x_i^2 - \left(\sum x_i\right)^2}}$$

If we want the standard deviation of the intercept, therefore, we must obtain it by this separate calculation from "Std Err of Y Est."

DISTANCE, m	TIME(min), s	TIME(max), s	sqrt(DISTANCE)	TIME(calc)
0.10	0.143	0.153	0.316	0.137
0.20	0.191	0.201	0.447	0.197
0.30	0.239	0.249	0.548	0.243
0.40	0.285	0.295	0.632	0.281
0.50	0.310	0.320	0.707	0.315
0.60	0.347	0.357	0.775	0.346
0.70	0.380	0.390	0.837	0.374
0.80	0.398	0.408	0.894	0.400

Regression Output:

Constant	-0.00656
Std Err of Y Est	0.005217
R Squared	0.997152
No. of Observations	8
Degrees of Freedom	6
X Coefficient(s)	0.454867834598
Std Err of Coef.	0.00992360903

(a)

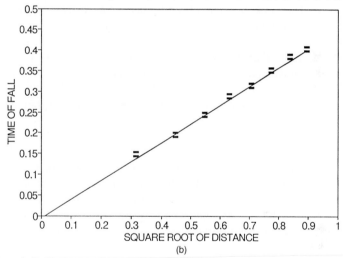

TIME vs. SQUARE ROOT OF DISTANCE

(b)

Figure 6–10 The least-squares analysis of the falling ball experiment using Quattro Pro.

The regression output also gives us (along with a few things that do not concern us) the value for the regression coefficient r. It is called "R Squared," and we can see that, as we would expect for a well-controlled experiment, its value is very close to 1.

After fitting a line to observations by using least squares, we usually want to illustrate the situation by using a graph that shows the observations and the least-squares line together, as shown in Figure 6–10(b). The graph was drawn in much the same way as was the one shown in Figure 6–9. The points were produced by exactly the same procedure, and the only difference lay in the way the line was produced. For Figure 6–9 we wished to plot an explicit model function, and we had to calculate the y values for that function. In Figure 6–10 we are pretending that we do not have an explicit model function, and we have used the least-squares procedure to generate a function. To plot the line, therefore, we must use the least-squares values of slope and intercept to calculate y values on that line. These values appear in Figure 6–10(b) in column E, and in the graphing process, that column was selected for the second graph and the "Lines" option was chosen for it. As before, such a diagram makes a suitable illustration for a report.

(3) Function Finding.   We have already discussed in Sections 6–10 and 6–11 the philosophy and logic of finding functions to fit observations for which no preexisting model is available. There is no need to repeat that discussion here except to emphasize again that, even if one does find an empirical function that matches the observations to a certain extent, there is no necessary significance in that function. It may be useful for interpolation or, if treated with *great* caution, extrapolation, but it cannot be claimed to supply anything more than that. It may turn out to be useful as a guide in building a model based on basic principles, but it also may not.

We have mentioned here (see also Appendix 3) some procedures that can be used to find suitable functions by means of hand-done calculations, tbut he methods are slow and tedious. Computers make the task easy, and one can quickly try various possible functions. Most of these programs fit their various functions by using the least-squares criterion, either directly and analytically, or else by some process of successive approximations that use repeated iteration. One such program is SlideWrite, which was used to create Figure 6–11 with its various curves fitted to the observations from Table 4–1.

SlideWrite has one additional advantage in that it will plot directly the uncertainty on the points, either in the vertical or horizontal direction (but not both together). Figure 6–11(a) shows the observations plotted directly. In Figure 6–11(b) we see the result of fitting a straight line to the observations, (c) shows an exponential function, (d) shows a logarthmic function, and (e) shows a power law (i.e., a function of the type ax^b). Clearly, none of these is a good fit to the observations, but if we had had some good reason to use such a function, the program would give us the best choice of parameters for that chosen function on the basis of the least-squares criterion. Figure 6–11(f) shows a better fit. It is for the *sigmoidal* function, which has the form

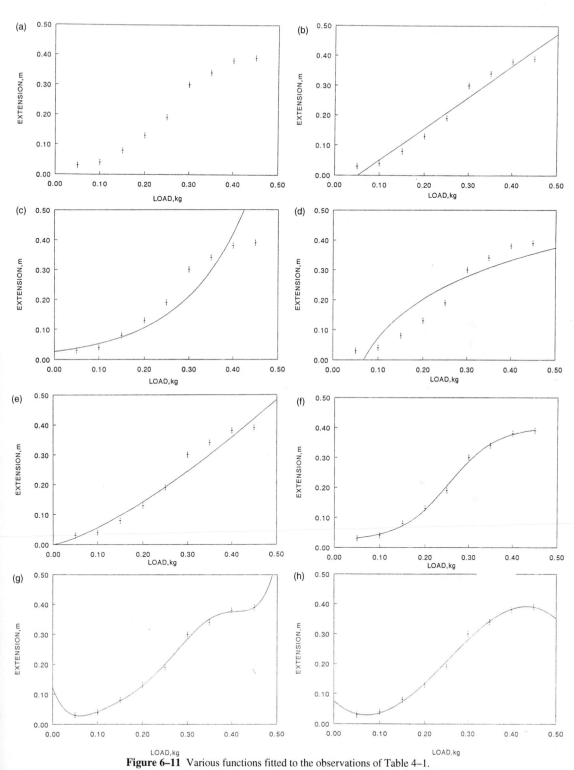

Figure 6–11 Various functions fitted to the observations of Table 4–1.

$$y = a_0 + \frac{a_1}{1 + \exp\left(-\dfrac{(x - a_2)}{a_3}\right)}$$

and the program has given us the values of the various a's optimized to meet the least-squares criterion using an iterative process.

Last, Figures 6–11(g) and 6–11(h) provide graphic illustration of an interesting and important point. Both represent polynomial fits to the observations, but they are calculated for two different orders in the polynomial function. In Figure 6–11(g) the polynomial chosen contained terms up to the third order—that is, the function was of the form

$$y = a_0 + a_1 x + a_2 x^2 + a_3 x^3$$

and the computer gave us the optimized values of the coefficients on the basis of least squares. Figure 6–11(h) shows the result of fitting a sixth order polynomial to the measurements. Both of these functions would obviously serve us well if we were interested in interpolation only (although it is apparent that the sixth order polynomial provides a slightly better fit than the other). Their behavior *outside* the measured range, however, provides convincing evidence that, as was pointed out in Section 4–2, extrapolation on the basis of a model that is wholly empirical (i.e., based on nothing other than the observations themselves), is a highly questionable exercise.

PROBLEMS

1. An experiment was done to measure the impedance of a series *R-L* circuit. The impedance Z is given as a function of the resistance R, the frequency of the source f and the inductance L by

    $$Z^2 = R^2 + 4\pi^2 f^2 L^2$$

    The experiment was done by measuring Z as a function of f with the intention of plotting Z^2 vertically and f^2 horizontally to obtain L from the slope and R from the intercept. The observations obtained are given in the table.

f, Hertz	Z, Ohms	f^2	$f \times \mathrm{AU}(f)$	$\mathrm{AU}(f^2)$ $= 2f \times \mathrm{AU}(f)$	Z^2	$Z \times \mathrm{AU}(Z)$	$\mathrm{AU}(Z^2)$ $= 2Z \times \mathrm{AU}(Z)$
123 ± 4	7.4 ± 0.2						
158	8.4						
194	9.1						
200	9.6						
229	10.3						
245	10.5						
269	11.4						
292	11.9						
296	12.2						

The uncertainties given in the first line refer to all the readings in each column.

(a)   Plot these readings in the appropriate fashion, and mark the uncertainties on the points. Suggested table headings to expedite the calculations are given above.

(b)   See if the observations can be interpreted in terms of a straight line for any part of the range or all of it.

(c)   Obtain the slope of the best line.

(d)   Calculate the best value for L.

(e)   Obtain the slopes of the lines at the outer limits of possibility, and so state the range of uncertainty for the slope.

(f)   Calculate the absolute uncertainty in the measurement of L.

(g)   Obtain the best value of R from the intercept.

(h)   Obtain the uncertainty for the R value.

(i)   State the complete result for the experiment with the appropriate number of significant figures in each quantity.

2.   Ten observers report on the intensity of a lamp measured repeatedly by using a comparison photometer. Their results (in arbitrary units) are as follows:

Observer	Mean of Intensity Measurements	Standard Deviation of Mean
1	17.3	2.1
2	18.4	1.9
3	17.1	2.5
4	16.6	2.8
5	19.1	3.2
6	17.4	1.2
7	18.5	1.8
8	14.3	4.5
9	16.8	2.3
10	17.4	1.6

What is the overall mean value for the intensity, and what is its standard deviation?

3. An experiment has been carried out to investigate the temperature dependence of the resistance of a copper wire. A common model is represented by the equation

$$R = R_0(1+\alpha T)$$

where R is the resistance at temperature $T°C$, R_0 is the resistance at $0°C$, and α is the temperature coefficient of resistance. The observations of R and T that were obtained follow:

$T°C$	R Ohms
10	12.3
20	12.9
30	13.6
40	13.8
50	14.5
60	15.1
70	15.2
80	15.9

Assume that the uncertainty in the measurements of temperature can be neglected.

(a) Using the method of least squares (i.e., using directly the equations of Section 6–7), evaluate the slope and intercept of the graph of R versus T.

(b) Hence, obtain the best value for α.

(c)   Evaluate the standard deviation for the slope and for the intercept.

(d)   Hence, evaluate the standard deviation of α.

(e)   State the final result of the experiment with the appropriate number of significant figures.

4. It is desired to fit a set of observations to the function $y = a + bx^2$ by using least squares. Use the same procedures that are used in Appendix 2, Section A2–2, for calculating the constants of a linear function to obtain equations for a and b in the parabolic function. Hence, calculate the values of a and b appropriate to the following set of observations:

x	y
0.5	1.5
1.0	6.3
1.5	12.4
2.0	12.6
2.5	18.0
3.0	32.8
3.5	40.2
4.0	47.4

Assume that uncertainty is confined to the y variable.

5. The following measurements were made in the investigation of phenomena for which no existing model was available. In each case identify a suitable function and evaluate its constants.

v	i	x	y	T	f
0.1	0.61	2	3.2	100	0.161
0.2	0.75	4	16.7	150	0.546
0.3	0.91	6	44.2	200	0.995
0.4	1.11	8	8.2	250	1.438
0.5	1.36	10	150.7	300	1.829
0.6	1.66	12	233.5	350	2.191
0.7	2.03	14	337.9	400	2.500
0.8	2.48	16	464.5	450	2.755
0.9	3.03	18	618.0	500	2.981
(a)		(b)		(c)	

6. The following results come from a study on the relationship between secondary-school matriculation averages and the students' overall aver-

ages at the end of first-year university. The first number of the pair is the secondary-school average and the second is the university average.

78,65; 80,60; 85,64; 77,59; 76,63; 83,59; 85,73; 74,58; 86,65; 80,56; 82,67; 81,66; 89,78; 88,68; 88,60; 93,84; 80,58; 77,61; 87,71; 80,66; 85,66; 87,76; 81,64; 77,65; 96,87; 76,59; 81,57; 84,73; 87,63; 74,58; 91,78; 92,77; 85,72; 86,61; 84,68; 82,66; 81,72; 91,74; 86,66; 90,68; 88,60.

(a) Draw a scatter diagram of university averages plotted against school averages.

(b) Evaluate the correlation coefficient.

7. Evaluate the correlation coefficient for the values of $\sqrt{x}$ and t in Table 4–3.

7

Writing Scientific Reports

7–1 GOOD WRITING DOES MATTER

It is almost impossible to overestimate the importance of good scientific writing. The best experimenting in the world can be of little or no value if it is not communicated to other people—and communicated well by clear and attractive writing. Although communication may sometimes be verbal, in the overwhelming majority of cases people learn about our work from the printed page. Our obligation to become as literate as we can is therefore not trivial, and it should be regarded as an essential and integral part of our experimenting activities. Our writing must be sufficiently good to attract and retain the interest and attention of our readers. This chapter contains some hints on how this may be achieved, and the sample report in Appendix 4 is an example of the ways in which the suggestions can be implemented in practice.

It is almost impossible to tell a person how to write well. It would be very convenient if we could lay out a short list of instructions and guarantee thereby fluent, lucid, and literate prose, but the list is not available. Each of us has different ways of expressing thoughts, and each must allow his or her writing style to develop in its own way. This needs extensive practice, and we should regard report writing in the introductory physics laboratory as an excellent opportunity to obtain it. We may end up acquiring different writing styles, but provided the message is clear, the diversity can be enriching rather than damaging.

We now turn to some practical considerations concerning the actual writing of reports. Although there is no list of explicit instructions for good writing, there is one principle that will make the production of good, readable prose more likely. Whether one is preparing a report for internal circulation in a private organization or a paper for publication in the open literature, there is one person whose interests must claim the writer's first attention—the person who will actually read the report. As far as the report is concerned, that person is the most important person in the world. We are well advised to concentrate our attention on him or her. Our readers are very likely people we do not know, perhaps in some far-distant part of the world and very likely knowing nothing about us or our work except the report that they hold in their hands. We probably have only one chance to influence them—as they read our report—and the report must do it alone. We cannot stand beside our readers, adding explanation and clarification if they encounter difficulty in understanding what we have written. Not only must the report stand on its own, the outcome of the reading can be highly significant. The public recognition of our work, the opportunity for others to benefit from it, our own reputation, perhaps even our chances of employment or promotion may depend on these few minutes spent by our readers as they work their way through our report. Do we have to be persuaded further that we should take writing seriously?

It has been common in the past to recommend a detached, depersonalized mode of expression characterized by the use of the passive voice and impersonal constructions. There seems to be little point in perpetuating such stilted language. We can simply tell our readers what we did in our experiment—for example: "We measured the time of fall using an electronic timer that was accurate to 1 millisecond." Since there is no single "right" way to write a report of an experiment, we should feel free to use such language and modes of expression as allow us to express our thoughts in the most clear, attractive, and persuasive way possible. For invaluable advice regarding writing style, consult the little book by Strunk and White that is listed in the Bibliography.

We now consider the various sections of the report in turn, all as seen through the eyes of our all-important reader.

7–2 TITLE

The title is probably the first part of the report to draw the attention of readers. Because they are almost always busy people with many items competing for their attention, our report will capture their interest only if the title is informative, appropriate, and attractive. It should not be too long, yet it should specify quite explicitly the topic of the work. For example, if the purpose of the experiment is to measure the specific heat of a fluid by using continuous-flow calorimetry, we can use this fact directly as a title: "Measurement of the

Specific Heat of Water by Using Continuous-Flow Calorimetry." Notice that three questions are answered in this title:

1. Is the work experimental or theoretical? That is, are we reporting a measurement or a calculation?
2. What is the topic of the work?
3. What general method did we use?

Attention to these three items will almost invariably result in a good choice of title.

7-3 FORMAT

The sections that follow analyze the various parts of a report. The various subsections that are described under each section heading should not themselves be used as headings in actual reports. Although practice obviously varies with circumstances, reporting on most normal work in the introductory physics laboratory needs only the minimum of division. The sections of a report that are essential are the following:

INTRODUCTION
PROCEDURE
RESULTS
DISCUSSION

These divisions can be used as a basic starting point. The headings should be neat, clear, and written in block capitals. Subsections within each of these main sections should be used only when the length or complexity of the report makes them indispensable for clarity. Other main sections may be introduced in accordance with the requirements of particular experiments. Suggested possibilities are the following:

THEORY
SAMPLE PREPARATION
UNCERTAINTY CALCULATIONS

To make the report as inviting to read and as easy to understand as possible, it should contain a clear, logical thread of argument, and we should not allow anything to disrupt that development of thought. If we feel that we must include some particular piece of description that is so lengthy and detailed that it would interrupt the smooth development of the main argument, we should consider making it an appendix to the report. In that way all the detail is available to any reader who wants it, but the main continuity of thought is not broken.

Let us turn now to the details of each section of the report.

7–4 INTRODUCTION

The various components that make up an informative introduction are, in order of presentation, as follows.

Topic Statement
Review of Existing Information
Application of Information to Specific Experiment
Summary of Experimental Intention

Topic Statement

With a good title, we can assume that we now have our readers' attention and that they have picked up the report. However, they are almost certainly starting from zero, or close to it, as far as our particular experiment is concerned. As they start to read, our first task is to orient their thinking toward the particular area of study. We are not going to succeed in this by diving immediately into unorganized detail about the experiment. Think instead of the most general statement that can be made about the experiment and state it directly. For example: "It is possible to measure gravitational acceleration by using the oscillation of a simple pendulum." In this way readers are taken from their initial state of ignorance to direct awareness of the specific topic of the work.

Review of Existing Information

At this point readers need some reminder of the basic information relating to this particular area. We can meet this need by giving them a brief summary of the existing state of knowledge relevant to the experiment. The summary may include, as necessary, some aspects of the history of the subject, a summary of earlier experimental work, or both. Two items are not discretionary and must be included in every report on an experiment. One is a clear statement of the system and the experimental circumstances with which the report deals; and the second is a description of the model or models used.

It is generally best to give this summary of existing information quite briefly for fear of obscuring the main line of argument, but it should be sufficiently detailed so that readers can understand the rest of the report. In the interests of brevity and clarity, the derivation of standard theoretical results associated with the model should not be included. (The way in which these standard results are manipulated to refer to our particular system is another matter, however, because that is specific to the experiment. This is the topic of later discussion.) The behavior of the model, as represented by important equations, should be quoted, and it is important at this stage to mention any assumptions contained in the model that may limit the validity of the equations. For example: "It can be proved that, in the limit of vanishingly small amplitude of oscillation, the period of oscillation of a simple pendulum, considered to be a point mass at the end of a massless, inextensible string, is given by ... ." To compensate for the omission of standard derivations it may be desirable to include in the references a source in which the complete derivation can be found.

Application of Information to Specific Experiment

Readers are now equipped to understand all that follows in the report, and their natural reaction at this point will be to wonder: "How does all this refer to this particular experiment?" We therefore supply a paragraph or two to show how the basic information, such as an equation representing the behavior of the model, can be converted to provide a foundation for our particular experiment. Commonly, this involves some procedure such as putting the basic equation into straight-line form (or some suitable equivalent) and identifying the ways in which the model can be tested against the system. We can also point out at this stage the information that will become available from the parameters of the graph (such as slope and intercept in the case of straight-line plotting). Readers thereby become fully aware of how our final answer will be obtained.

Summary of Experimental Intention

It is helpful to readers to conclude the introduction with a summary of our specific intention in the experiment. For example: "Thus, by measuring the variation of index of refraction with wavelength, we shall be able to test Cauchy's model using a graph of n vs. $1/\lambda^2$; the values of Cauchy's coefficients A and B that are appropriate to our glass specimen will then be obtained, respectively, from the intercept and slope of the graph." Such a statement is satisfying to the readers because, particularly in a long and complicated experiment report with a lengthy introduction, it offers them a review in summary form of the whole course of the experiment, and it enables them to make sense of the subsequent description of the actual conduct of the experiment.

Statement of Experimental Purpose

No mention has yet been made of the traditional statement of purpose for the experiment. It has been omitted so far because, although it should appear somewhere in the introduction, there is no universally suitable location. If the topic of the experiment is familiar, the statement of purpose can form an acceptable topic statement right at the beginning of the introduction. For example: "It is the purpose of this experiment to measure the acceleration of gravity by timing the fall of a freely falling object." Under suitable circumstances such a statement of purpose can make an excellent topic statement. On the other hand, the basic purpose of an experiment might involve matters so complicated and unfamiliar that a statement of it would be completely incomprehensible unless it followed a substantial amount of preparatory material. It is easy to imagine a complicated theoretical description that could profitably conclude with the phrase: "... and it is the purpose of this experiment to determine a value for the coefficient k in equation 10." It does not matter a great deal where the statement of purpose comes, so long as it is included and comes at a point in the introduction where it fits well and makes good sense to the readers.

The introduction has performed a number of services for our readers. Right at the beginning the topic statement has directed their attention to our particular area of work. They have then been reminded of the existing state of knowledge in that area. Next they have been shown how that applies to our particular experiment. Finally, they have been given a concluding summary of our specific experimental intention. They are now ready to hear how we actually did the experiment.

7-5 PROCEDURE

The report's introductory section takes the form of a descriptive sequence that proceeds from *general* to *specific*. We start with a topic statement that is the most general remark about the experiment we can make, and we end with a completely specific statement of intention. Such a sequence is designed to suit readers' requirements in the introductory section, and a similar sequence is equally suitable for the procedure section. If we were to start the description of procedure by launching immediately into a mass of unorganized detail, we would succeed only in irritating our readers. How can they appreciate the great care we took in some experimental detail if they are not aware even of our choice of measured variables? We should be as considerate of our readers' mental efforts when we write the procedure section as we were in the introduction. A second sequence from general to specific is clearly called for.

Outline of Procedure

To set the scene for the subsequent description of the details of procedure and measurement, we first offer readers a review of the whole course of the experiment. If the experiment really consisted of the measurement of the variation of electrical resistance of a copper wire with temperature over the range 20°C to 100°C, we say just that to provide the readers with a frame-work into which they can fit all subsequent description of detail. If we start the description of procedure by saying that we connected terminal A to ter-minal B, switched on power supply C, read voltmeter D, and the like, we shall have lost their attention in two lines.

Specific Measurement Details

Now that the readers know the general course of the experiment, they are ready to be told the specific methods by which we measured each of the re-quired quantities, carried out sample preparation, and performed other steps. This can be done quite simply by stating each in turn until we have com-pleted the list. We must make sure that no significant method of measure-ment is omitted; in such a thing as a timing measurement it is almost cer-tainly important that we used an electronic timer with millisecond accuracy rather than a stopwatch that could be read to 0.2 second, and our readers need to know that we did so. If a quantity in the experiment could be measured by using some standard and familiar technique, it may be sufficient to mention it by name. For example: "The resistances were measured using a Wheatstone bridge accurate to 0.01%." If we feel it is unusually significant, we can discuss at this stage the accuracy of any particular measuring process, while remembering that the overall precision of the experiment is a different topic that will appear in a subsequent section of the report.

Precautions

After readers have learned the methods by which we made each measure-ment, they may recall the difficulties or possibilities for error that are inher-ent in particular procedures. They therefore need reassurance that we, too, had thought of these possibilities and had been sufficiently careful to take the necessary precautions. As we offer that reassurance, however, we need not go to extremes. Care should obviously be taken with *all* measurements; there is no point in making superfluous claims to virtue in describing routine and obvious precautions. There are times, however, when special care to avoid some particular source of error is a genuinely important part of the experiment, and it is reasonable to draw attention to this before we close the procedure section.

Apparatus Diagrams

Good diagrams of experimental apparatus are an essential part of any good report. Although a published paper requires drawings of professional quality, such resources are not available in introductory work. However, we should early acquire the habit of taking care with drawings of apparatus. Even if sophisticated drawing aids are not available, it is not too much to expect the use of a ruler. Neatness and clarity are important, and good, legible labeling assists enormously in understanding the experiment. Good diagrams can also help experimenters to write reports. Reference to a good, clear, well-labeled diagram can save paragraphs of written description and provide detail that would be intolerably tedious to read if it were included in the text.

Reference to diagrams can be inserted at any appropriate point in the text, but reference to a general diagram of the apparatus as a whole can make a neat and convenient beginning for a procedure section. For example: "Using the apparatus shown in Figure 1, we measured the variation of time of fall of a ball bearing with height over the range 20 cm to 150 cm." Figure 7–1 is an example of an acceptable apparatus diagram.

7–6 RESULTS

Measured Values

At this point in the report, readers have all the information they need for understanding the experiment, and they are ready to receive the results directly. Because any good experiment almost inevitably involves the variation of some quantity with another, the results are usually best presented in a table. As always, neatness and clarity are of paramount importance. Lines for the table should be drawn with a ruler, and ample space should be provided for the headings and for the columns of figures. The headings should be explicit and should include, if possible, the name of the variable, its symbol, and the units of measurement. Attached to each numerical entry should be its uncertainty, unless some separate discussion of uncertainties makes the precision of the measurements absolutely clear. Tables should be clearly identified

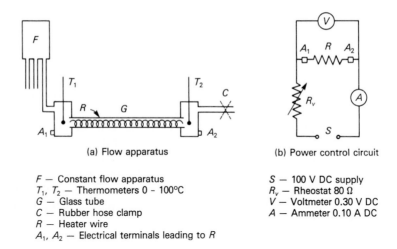

(a) Flow apparatus   (b) Power control circuit

F — Constant flow apparatus   *S* — 100 V DC supply
T_1, T_2 — Thermometers 0 - 100°C R_v — Rheostat 80 Ω
G — Glass tube   *V* — Voltmeter 0.30 V DC
C — Rubber hose clamp   *A* — Ammeter 0.10 A DC
R — Heater wire
A_1, A_2 — Electrical terminals leading to *R*

Figure 7–1 A completed apparatus diagram.

with a table number and a title. At this point it may be appropriate to refer to any graphs we have drawn of the basic variables. A simple statement will suffice. For example: "A graph of the time of fall versus height is shown in Figure 2. Any tables of values that are so extensive and detailed that their inclusion in the main text of the report would disrupt the readers' train of thought should be relegated to an appendix.

Following the main table or tables, list the measured values of all the other quantities in the experiment. As always, each should have its uncertainty attached, and the units of measurement should be clearly stated.

Description of Measurement Uncertainties

The report should state explicitly the kind of uncertainties we are quoting. These are likely to be either estimated outer limits or statistical quantities such as a standard deviation or a standard deviation of the mean. In the case of statistical quantities, we must not omit mention of the number of readings in the sample from which the results were derived. If any quantities in our list of measured values were obtained by computation from some basic measurement or measurements, we must state clearly the type of calculation used to obtain the final uncertainty in the computed quantity. It is not necessary to give much, if any, of the arithmetic details, provided that our readers can see clearly the kind of calculation that was performed.

Computation of Final Answer

If the experiment has been well designed, we will probably obtain our final answer by some graphical procedure. It is now time to tell the readers exactly what that procedure is. In simple cases we may obtain the answer from the basic graph of one measured variable against the other, but even then we must be explicit about what we have done. For example: "The value of the resistance was obtained from the slope of the graph (shown in Figure 3) of V versus I between 0.5 A and 1.5 A." If the answer was not the slope itself but was obtained from calculation with other measured quantities, we must state explicitly what we have done. For example: "Our value for the coefficient of viscosity was obtained from the slope of the graph of Q versus P in combination with the measured values of a and ℓ using Equation (3)."

The readers will wonder what kind of calculation we performed to obtain the uncertainty in the final answer. We simply say what we have done. If we assessed visually the possible range of slopes, we say just that. We can add, if necessary, that the basic uncertainty in slope was combined with other uncertainties, and state explicitly the method of calculation. If we obtained the slope by a least-squares calculation and incorporated any other standard deviations to obtain a final value for the uncertainty of the answer, we again state simply what we have done.

Throughout the results section of the report, we do not trouble our busy readers with unnecessarily detailed calculations. They trust us to do simple arithmetic, but they want to know what kind of calculation we did. If we feel compelled, for some particular reason, to offer an unusual amount of detail regarding such calculations, we can always put it in an appendix where it will be available if wanted but where it will not obscure the clarity of the main report.

7–7 GRAPHS

Graphs in the report differ from the graphs used in doing the experiment. Those graphs were working documents designed as computational aids. For a precise experiment, the graphs are possibly quite large and finely drawn to permit precise extraction of information. On the other hand, it is extremely unlikely that our readers will want to do any numerical work of their own using the graphs in the report. These graphs serve mostly as illustrations. They allow the readers to see the behavior of the system so that they can judge for themselves the validity of our assertions about the results.

The graphs in the report must be clear, neat, and uncluttered so that readers do not have to work too hard to get the message. The points on the graphs should have their uncertainties clearly marked on them (by a box or a cross), and the axes should be clearly labeled. Both the type of uncertainty and any symbols used in labeling the axes should be explicitly identified in some obvious way in or beside the graph; we do not want to cost our readers the irritation of hunting through the text to find out how to interpret the graph. Do not, however, fill up empty spaces on the graph with arithmetic calculations of slopes, and the like. Each graph should obviously have a clear title or, as is common in printed publications, a more extended caption. In addition to supplying identification, an extended caption has the added advantage of supplying a good location for the important details mentioned earlier. A sample of acceptable layout for such an illustrative graph is given in Figure 7–2.

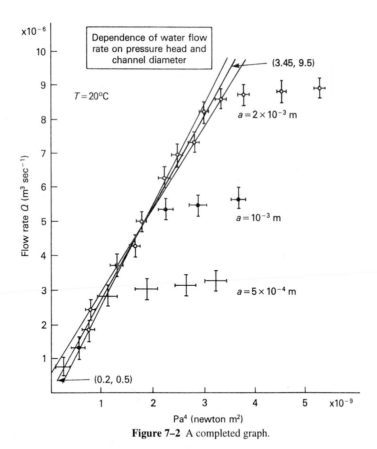

Figure 7–2 A completed graph.

7–8 DISCUSSION

Comparison Between Model and System

The discussion is an integral part of the report and not an afterthought. It has this importance because we have yet to describe the aspect of the experiment that we have, right from the beginning, regarded as the basic issue in experimenting—the relationship between the system and the model. The outcome of that comparison is vital for the experiment, and our readers will be eager to hear what we have to say about it.

We have already listed in Section 6–11 the various categories of experimental outcome. Remembering that, in evaluating our results, we had to disengage ourselves from our hopes and aspirations for the experiment and accept objectively the actual outcome, so now, at the reporting stage, we must make a candid and unbiased statement of that outcome. We should make it a plain, simple statement of the actual situation. For example: "The behavior of the model is represented by Equation (1), in which the variation of Q with P is a straight line passing through the origin. In the experiment the results do show a linear variation over part of the range, but instead of passing through the origin, the line that best fits the observations has a finite intercept on the Q axis. In addition, at the high end of the P scale, the observations show systematic deviation from linearity by an amount clearly in excess of the uncertainty of measurement."

At this stage it is sufficient to make this plain, factual statement. Because such comparison was the fundamental objective of the experiment, it is necessary for its outcome to be clearly, factually, and prominently stated. In the report we shall be proceeding quite soon to matters of interpretation and opinion, and it is important that we start the discussion section with a plain statement of the actual, indisputable outcome of the experiment.

That statement will raise some questions in the mind of the readers, and we must turn our attention to them now.

Consequences of Discrepancies Between Model and System

One of these questions concerns the possibility of error in the final answer that could be caused by failure of correspondence between the system and the model. Some of these possibilities have already been mentioned in Section 4–5, and readers need reassurance that we have protected the final answer from that kind of error. We should point out, for example, that an unexpected intercept will not contribute to error in a quantity that has been obtained from the slope alone, or that a systematic departure from linearity over part of a graph did not invalidate an answer that was obtained from the linear segment only. Much of the skill in experimenting lies in the protection of the final answers from such sources of error, and we can be quite explicit about our claims to have done so.

Speculation Concerning Discrepancies Between System and Model

In describing the report's earlier sections, we have stressed objective and factual reporting of the actual situation. Matters of opinion or conjecture should not have played a significant role in those parts of the report, and we have probably limited ourselves to such statements as would have been made by most impartial observers. Now, however, comes a stage at which we not only can but should introduce our own ideas. Our readers have in turn been informed about the actual degree of correspondence between the system and the model, and they have been reassured that the final answer has not been contaminated (as far as we were able to tell) by any failure of correspondence between the system and the model. Because we have met our basic obligations as experimenters, we could quite justifiably leave the report there. However, the interest of the readers will doubtless have been aroused by the description of any discrepancies between the model and the system. We presumably started with a model that was chosen to suit the system as closely as possible. If any failure of correspondence between the system and the model had been anticipated, such breakdown would have been incorporated into the experiment design. If we want to measure a coefficient of viscosity by using theories based on streamline flow, for example, we do not design an experiment to run under turbulent conditions (unless, for some separate purpose, we wish to detect the onset of turbulence). Any observed failure of correspondence, therefore, is bound to attract attention, and readers will want to know what we think about it. We are more familiar with the experiment than anyone else and are in a better position than others to guess at the origin of discrepancies.

Sometimes a discrepancy has (at least superficially) an origin that is easy to identify. If we measure, for example, the flow rate of fluid through a pipe, a departure of the flow-rate measurements below linear behavior at high values of pressure difference may be ascribed quite confidently to the onset of turbulence. If the objective of the experiment included the detection of the onset of turbulence, such a statement could end the matter. At other times, however, more comment is needed. If, in the preceding example, our intention had been simply to measure the coefficient of viscosity from the linear part of the Q, P variation, readers might wonder why we had not been more successful in avoiding a region in which the streamline theory was clearly invalid. Perhaps we had been surprised by an unexpectedly early on-set of turbulence; if so, we should be candid enough to admit it and perhaps speculate on the origin of that discrepancy. If the situation is genuinely puzzling, we may not be able to offer much in the way of speculation, but it is always worth trying. As has been said, as the experimenters, we have a better chance of speculating fruitfully than most others, and our ideas are al-most certainly to be of interest and possible value to other workers.

Sometimes, however, despite our best efforts we fail and are unable to offer any constructive ideas. We must be completely honest. If we are dealing with a well-tested system and a well-known, reliable model and if we have tried and failed to resolve a failure of correspondence between them, our situation cannot but be of interest to other workers. We should tell them about it, and perhaps we shall all learn something from the resulting discussion.

As we attempt to be creative regarding our experimental discrepancies, we should remember that we are doing something important. All models and theories go through processes of refinement, and these processes are based on the various stages of observed failure of the models. We should try to be responsible, therefore, as we speculate. Instead of having a fling at every wild idea we can imagine, we should try to make our suggestions have some logical connection with the evidence of the discrepancy. For example, if we have done an experiment on the oscillation of a load suspended from a spring, we could write: "Since the unexpected intercept in the plot of T^2 vs. m gives a finite value for T at $m = 0$, we have a clear indication of the pres-ence of an extra mass that was not included in the measured values of load." Whether we can guess at the *identity* of this extra mass is less important. We have at least offered a logically acceptable inference from the observed na-ture of the discrepancy, and further research and experimenting in this area will have been facilitated.

Appendix 1

Mathematical Properties of the Gaussian or Normal Distribution

A1–1 THE EQUATION OF THE GAUSSIAN DISTRIBUTION CURVE

To derive the equation of the Gaussian distribution curve, we consider a quantity, whose unperturbed value is X, to be subject to random uncertainty. We assume that the uncertainty arises from a number, $2n$, of fluctuations from the central value, each of magnitude E and equally likely to be positive or negative. The measured value x can then range all the way from $X - 2nE$, if all the fluctuations have the same sign in the negative direction, to $X + 2nE$ if the same thing happened positively. Intermediate values arise from various combinations of positive and negative contributions. We wish to determine the form of the resulting distribution curve for a very large number of such measurements. This form will be determined by the probability of encountering a particular deviation R within the total interval $\pm 2nE$. The probability is governed by the number of ways in which a particular deviation can be generated.

For example, a deviation of the total value $2nE$ can be generated in only one way: All the elementary contributions to the deviation must have the same sign simultaneously. An error of magnitude $(2n - 2)E$, on the other hand, can occur in many ways. If any one of the elementary fluctuations had been negative, the total deviation would have added up to $(2n - 2)E$, and this

171

situation can arise in $2n$ different ways. A deviation of $(2n - 2)E$ is, therefore, $2n$ times as likely to happen as one of $2nE$. A situation in which two of the elementary fluctuations have negative signs can, correspondingly, be generated in many more ways than for one, and so on.

The argument can be generalized by using the number of ways in which a specific deviation R can be generated as a measure of the probability of the occurrence of that deviation and, consequently, as a measure of the frequency of its occurrence in a universe of observations.

Consider a total deviation R of magnitude $2rE$ (where $r \leq n$). This must be the result of some combination of fluctuations of which $(n+r)$ are positive and $(n-r)$ are negative. The number of ways in which this can happen can be calculated as follows. The number of ways of selecting any particular arrangement of $2n$ things is $(2n)!$. However, not all these arrangements are different for our purpose, because we do not care if there is any internal rearrangement within the fluctuations in, say, the positive group. We must, therefore, divide the total number of arrangements by the number of these insignificant rearrangements, that is, by $(n + r)!$. Similarly we must divide by the number of internal rearrangements that are possible in the negative group, [i.e., by $(n - r)!$]. The total number of significant combinations is, therefore,

$$\frac{(2n)!}{(n+r)!(n-r)!}$$

This quantity is not yet strictly a probability, although it is a measure of the likelihood of finding such a total deviation. The probability itself is obtained by multiplying the foregoing number by the probability of this combination of $(n+r)$ positive and $(n-r)$ negative choices. Because the probability of each choice is $1/2$, the required multiplier is

$$\left(\frac{1}{2}\right)^{(n+r)}\left(\frac{1}{2}\right)^{(n-r)}$$

The final result for the probability of the deviation R is then

$$\frac{(2n)!}{(n+r)!(n-r)!}\left(\frac{1}{2}\right)^{(n+r)}\left(\frac{1}{2}\right)^{(n-r)} \tag{A1–1}$$

Our problem now is to evaluate this result as a function of the variable r. This is done subject to the condition that n is very large, tending to infinity. The evaluation requires two auxiliary results.

1. The first auxiliary result:

$$n! \approx \sqrt{2\pi n} e^{-n} n^n$$

This is known as Stirling's theorem. Although its full derivation is beyond our scope, its plausibility can be indicated as follows.

Note first that

$$\int_1^n \log x \, dx = \left[x \log x - x \right]_1^n$$

$$= n \log n - n + 1$$

Now the graph of $\log x$ versus x is shown in Figure A1–1, and it shows clearly that the value of the integral $\int_1^n \log x \, dx$ can be approximated by the sum

$$\log 1 + \log 2 + \log 3 + \ldots + \log n$$

which is $\log(1 \times 2 \times 3 \times \ldots n)$ or $\log n!$. We can, therefore, write approximately, if n is large,

$$\log n! = n \log n - n$$

or

$$n! = e^{-n} n^n$$

This is an approximation to the formula given above. A full derivation is in the text by Margenau and Murphy listed in the Bibliography.

2. The second auxiliary result is:

$$\lim_{n \to \infty} \left(1 + \frac{1}{n} \right)^n = e$$

This can be proved as follows. The expansion for $\left[1 + (1/n) \right]^n$ is

$$1 + \frac{n}{1!} \frac{1}{n} + \frac{n(n-1)}{2!} \left(\frac{1}{n} \right)^2 + \frac{n(n-1)(n-2)}{3!} \left(\frac{1}{n} \right)^3 + \ldots$$

As n becomes larger, all the terms involving n clearly tend to unity, so that the series tends to

$$1 + \frac{1}{1!} + \frac{1}{2!} + \frac{1}{3!} + \ldots$$

and the sum of this series to infinity has the value e, which gives us the required result.

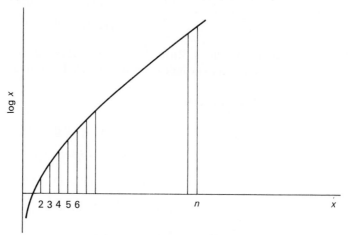

Figure A1–1 Graph of log x versus x

We are now in a position to evaluate the expression (A1–1). We apply Stirling's theorem to the terms $(2n)!$, $(n+r)!$, and $(n-r)!$. We obtain

$$(2n)! = (2n)^{2n} e^{-2n} \sqrt{2\pi \times 2n} = 2^{2n} n^{2n+(1/2)} e^{-2n} \sqrt{4\pi}$$

$$(n+r)! = (n+r)^{n+r} e^{-(n+r)} \sqrt{2\pi(n+r)}$$

$$= n^{n+r+(1/2)} \left(1+\frac{r}{n}\right)^{n+r+(1/2)} e^{-n-r} \sqrt{2\pi}$$

and

$$(n-r)! = n^{n-r+(1/2)} \left(1-\frac{r}{n}\right)^{n-r+(1/2)} e^{-n+r} \sqrt{2\pi}$$

Therefore,

$$(n+r)!(n-r)! = n^{2n+1} \left(1-\frac{r^2}{n^2}\right)^{n+(1/2)} \left(1+\frac{r}{n}\right)^{r} \left(1-\frac{r}{n}\right)^{-r} e^{-2n} \times 2\pi$$

The variable part of (A1–1) can now be written

$$\left(1-\frac{r^2}{n^2}\right)^{-n-(1/2)} \left(1-\frac{r}{n}\right)^{r} \left(1+\frac{r}{n}\right)^{-r}$$

$$= \left(1-\frac{r^2}{n^2}\right)^{-(n^2/r^2)(r^2/n)} \left(1-\frac{r^2}{n^2}\right)^{-1/2} \left(1+\frac{r}{n}\right)^{n/r(-r^2/n)} \left(1-\frac{r}{n}\right)^{-n/r(-r^2/n)}$$

Thus expression (A1–1) can now be written

$$\frac{1}{\sqrt{n\pi}}\left(1-\frac{r^2}{n^2}\right)^{-(n^2/r^2)(r^2/n)}\left(1-\frac{r^2}{n^2}\right)^{-1/2}\left(1+\frac{r}{n}\right)^{n/r(-r^2/n)}\left(1-\frac{r}{n}\right)^{-n/r(-r^2/n)}$$

Now, as $\dfrac{n}{r}\to\infty$, we have

$$\left(1-\frac{r^2}{n^2}\right)^{-n^2/r^2}\to e$$

$$\left(1-\frac{r^2}{n^2}\right)^{-1/2}\to 1$$

$$\left(1+\frac{r}{n}\right)^{n/r}\to e$$

$$\left(1-\frac{r}{n}\right)^{-n/r}\to e$$

and so, finally, the probability of deviation R is

$$\frac{1}{\sqrt{\pi n}}e^{-r^2/n}$$

The significant feature of this result is the form e^{-r^2}. It specifies the probability of a deviation R and is thus equivalent to Equation (3–3) in which the deviation is the difference between the unperturbed value X and the measured value x. The only problem that remains in putting the equation into standard form is to redefine the constants. Put

$$hx = \frac{r}{\sqrt{n}}$$

for the value of the exponent, and in the constant replace $1/\sqrt{n}$ by $h\,dx$. The equation then reads

$$P(x)\,dx = \frac{h}{\sqrt{\pi}}e^{-h^2x^2}\,dx$$

where $P(x)\,dx$ is the probability of finding a deviation between x and $x+dx$.

A1–2   STANDARD DEVIATION OF THE GAUSSIAN DISTRIBUTION

To calculate the standard deviation for the Gaussian distribution, we must calculate the standard deviation of a very large set of values that are distributed in a Gaussian distribution. We must, therefore, calculate the sum of the squares of the deviations from the central value and divide it by the total number of observations. Let there be N observations, where N can be assumed to be a very large number. The number of deviations from the central value that occur between x and $x + dx$ equals

$$\frac{Nh}{\sqrt{\pi}} e^{-h^2 x^2} dx$$

The value of the standard deviation, therefore, is given by

$$\sigma^2 = \frac{1}{N} \int_{-\infty}^{\infty} \frac{Nh}{\sqrt{\pi}} e^{-h^2 x^2} \times x^2 dx$$

$$= \frac{h}{\sqrt{\pi}} \int_{-\infty}^{\infty} x^2 e^{-h^2 x^2} dx$$

The integral is a standard one and has a value $\sqrt{\pi}/2h^3$, and so the value of the standard deviation is

$$\sigma^2 = \frac{h}{\sqrt{\pi}} \frac{\sqrt{\pi}}{2h^3} = \frac{1}{2h^2}$$

This provides the justification for Equation (3–4) and also enables us to rewrite the probability function $P(x)$ in terms of the standard deviation as

$$P(x)dx = \frac{1}{\sqrt{2\pi}\sigma} e^{-x^2/2\sigma^2} dx$$

The equation is frequently used in this form.

A1–3   AREAS UNDER THE GAUSSIAN DISTRIBUTION CURVE

It is frequently desirable to know what fraction of the area under the Gaussian distribution is enclosed within certain limits on the horizontal scale because this tells us the probability that observations will occur within that interval. To calculate these areas we proceed as follows. The probability that a deviation falls between x and $x + dx$

$$\frac{1}{\sqrt{2\pi}\sigma} e^{-x^2/2\sigma^2} dx$$

Therefore, the probability that a deviation lies between 0 and x is

$$\int_0^x \frac{1}{\sqrt{2\pi}\sigma} e^{-x^2/2\sigma}\,dx$$

Although this integral can be easily evaluated for infinite limits, it is not so simple for fixed limits as we now require. Numerical methods of integration are used, with results that are given in Table A1–1 (see also Figure A1–2).

TABLE A1–1 Areas under the Gaussian Curve

x/σ	Probability that a deviation lies between 0 and x
0	0.0
0.1	0.04
0.2	0.08
0.3	0.12
0.4	0.16
0.5	0.19
0.6	0.23
0.7	0.26
0.8	0.29
0.9	0.32
1.0	0.34
1.1	0.36
1.2	0.38
1.3	0.40
1.4	0.42
1.5	0.43
1.6	0.45
1.7	0.46
1.8	0.46
1.9	0.47
2.0	0.48
2.5	0.49
3.0	0.499

If we require the probability that a deviation lies somewhere within the complete range on *both* sides of the origin, that is, $\pm(x/\sigma)$, the value for the area is doubled. For example, the entry at $x/\sigma = 1$ is 0.34, which gives the 68% figure that we have been using for $\pm 1\sigma$ limits. The table gives only an indication of the way the probabilities run, and for serious statistical work reference should be made to one of the many statistical tables available (see the Bibliography under Lindley and Miller).

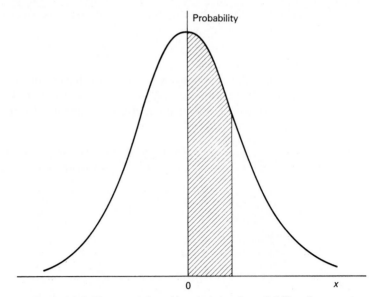

Figure A1–2 The area evaluated in calculating the probability of occurrence of an error up to x.

Appendix 2

The Principle of Least Squares

A2–1 LEAST SQUARES AND SAMPLE MEANS

Consider that we make N measurements, x_i, of a quantity that contains random fluctuation. Let us calculate that value, X, whose deviations from the x_i are minimized in accordance with the principle of least squares. X will be obtained from the condition

$$\sum (x_i - X)^2 = \text{minimum}$$

Let $\bar{x}$ be the mean of the x_i. Then

$$\sum (x_i - X)^2 = \sum \left[(x_i - \bar{x}) + (\bar{x} - X)\right]^2$$

$$= \sum \left[(x_i - \bar{x})^2 + (\bar{x} - X)^2 + 2(x_i - \bar{x})(\bar{x} - X)\right]$$

or, since $\sum (x_i - \bar{x}) = 0$,

$$\sum (x_i - X)^2 = \sum \left[(x_i - \bar{x})^2 + (\bar{x} - X)^2\right]$$

This last expression clearly has a minimum value when $\bar{x} = X$, thus confirming that the use of the mean as the most probable value for a sample is consistent with the principle of least squares.

A2–2 LEAST-SQUARES FITTED TO STRAIGHT LINES

Consider a set of observations (x_i, y_i) to which we wish to fit a linear relation

$$y = mx + b$$

We assume that the x values are precise, that all the uncertainty is contained in the y values, and that the weights of the y values are equal. (For the definition of the concept of "weights," see Appendix A2–3.) The deviations of the observed points from the straight line $y = mx + b$ are of the form

$$\delta y_i = y_i - (mx_i + b)$$

and we wish to have a minimum value for the sum of the squares of these quantities. We have

$$(\delta y_i)^2 = \left[y_i - (mx_i + b) \right]^2$$
$$= y_i^2 + m^2 x_i^2 + b^2 + 2mx_i b - 2mx_i y_i - 2y_i b$$

If there are N pairs of observations, the sum of the squares, M, is given by

$$M = \sum (\delta y_i)^2$$
$$= \sum y_i^2 + m^2 \sum x_i^2 + Nb^2 + 2mb \sum x_i - 2m \sum x_i y_i - 2b \sum y_i$$

The condition for the best choice of m and b is that $\sum (\delta y_i)^2$ should be a minimum. We need, therefore,

$$\frac{\partial M}{\partial m} = 0 \quad \text{and} \quad \frac{\partial M}{\partial b} = 0$$

The first condition gives

$$2m \sum x_i^2 + 2b \sum x_i - 2 \sum (x_i y_i) = 0$$

and the second gives

$$2Nb + 2m \sum x_i - 2 \sum y_i = 0$$

Solution of the two simultaneous equations for m and b gives

$$m = \frac{N \sum (x_i y_i) - \sum x_i \sum y_i}{N \sum x_i^2 - \left(\sum x_i \right)^2}$$

and

$$b = \frac{\sum x_i^2 \sum y_i - \sum x_i \sum (x_i y_i)}{N \sum x_i^2 - \left(\sum x_i\right)^2}$$

Standard deviations for m and b can be calculated as follows. Consider m first. Because m is a computed value that is calculated in terms of the uncertain quantities y_1, y_2, and the like, we can apply the equation that we had earlier (Equation 3–9) for the standard deviation of a computed value, z, that is a function of variables, x, y, and so on. It was

$$S_z^2 = \left(\frac{\partial f}{\partial x}\right)^2 S_x^2 + \left(\frac{\partial f}{\partial y}\right)^2 S_y^2 + \ldots$$

We apply this result to our case by noting that the x and y of the formula are the y_1, y_2, and so on that appear in the expression for m. We can write, therefore,

$$S_m^2 = \left(\frac{\partial m}{\partial y_1}\right)^2 S_{y_1}^2 + \left(\frac{\partial m}{\partial y_2}\right)^2 S_{y_2}^2 + \ldots$$

Now, in making our set of measurements of the x and y values, we would not normally have measured explicitly the standard deviation for each y value. In the absence of these we assume that the values of the various S_y's can be replaced by a quantity based on the scatter of the y values about the line whose m and b values we have calculated. These intervals, δy_i, have a standard deviation whose value (Equation 6–5) was

$$S_y = \sqrt{\frac{\sum (\delta y_i)^2}{N - 2}}$$

and this is the value that we shall use in place of all the separate S_{y_1}, S_{y_2}, and so on. Justification of the term $N - 2$ is not attempted here. It is associated with the fact that the δy_i are not independent but are connected by the existence of the best line that is specified by the values of m and b. The equation for the standard deviation of the slope can therefore be written

$$S_m^2 = \left(\frac{\partial m}{\partial y_1}\right)^2 S_y^2 + \left(\frac{\partial m}{\partial y_2}\right)^2 S_y^2 + \ldots$$

or

$$S_m^2 = S_y^2 \left[\left(\frac{\partial m}{\partial y_1}\right)^2 + \left(\frac{\partial m}{\partial y_2}\right)^2 + \ldots\right]$$

Let us write

$$S_m^2 = S_y^2 \sum_k \left(\frac{\partial m}{\partial y_k} \right)^2$$

where we have introduced a second index k to denote terms in the summation for the standard deviation. We must now evaluate the partial derivatives $\dfrac{\partial m}{\partial y_k}$.

The value of m is given by

$$m = \frac{1}{N \sum x_i^2 - \left(\sum x_i \right)^2} \left[N x_1 y_1 - y_1 \sum x_i + N x_2 y_2 - y_2 \sum x_i + \ldots \right]$$

and we can see that differentiation with respect to y gives us, for y_1 as an example,

$$\frac{\partial m}{\partial y_1} = \frac{1}{N \sum x_i^2 - \left(\sum x_i \right)^2} \left[N x_1 - \sum x_i \right]$$

or in general, for the k'th term,

$$\frac{\partial m}{\partial y_k} = \frac{1}{N \sum x_i^2 - \left(\sum x_i \right)^2} \left[N x_k - \sum x_i \right]$$

and

$$\left(\frac{\partial m}{\partial y_k} \right)^2 = \frac{1}{\left[N \sum x_i^2 - \left(\sum x_i \right)^2 \right]^2} \left[N^2 x_k^2 + \left(\sum x_i \right)^2 - 2 N x_k \sum x_i \right]$$

To obtain the value of S_m this must now be summed over the index k. If we write

$$\sum_k \left(\frac{\partial m}{\partial y_k} \right)^2 = \frac{1}{\left[N \sum x_i^2 - \left(\sum x_i \right)^2 \right]^2} \sum_k \left[N^2 x_k^2 + \left(\sum x_i \right)^2 - 2 N x_k \sum x_i \right]$$

and remember that $\sum_k x_k$ is the same thing as $\sum_i x_i$, we can easily perform the summation over the index k to obtain

$$\sum_k \left(\frac{\partial m}{\partial y_k}\right)^2 = \frac{1}{\left[N\sum x_i^2 - \left(\sum x_i\right)^2\right]^2}\left[N^2\sum x_i^2 + N\left(\sum x_i\right)^2 - 2N\left(\sum x_i\right)^2\right]$$

$$= \frac{1}{\left[N\sum x_i^2 - \left(\sum x_i\right)^2\right]^2}\left[N^2\sum x_i^2 - N\left(\sum x_i\right)^2\right]$$

$$= \frac{N}{N\sum x_i^2 - \left(\sum x_i\right)^2}$$

so that, finally, the value of S_m is given by

$$S_m = S_y \times \sqrt{\frac{N}{N\sum x_i^2 - \left(\sum x_i\right)^2}}$$

The value for S_b can be found by using the same procedure.

A2–3   WEIGHTING IN STATISTICAL CALCULATIONS

When we perform some statistical calculation, such as obtaining the mean of a set of observations or fitting a function to observations by using the least-squares principle, the equations in Sections 3–3 and 6–7 are valid only when all the observations are equally precise. If the measurements are of unequal precision, it is obviously fallacious to allow them to make equal contributions toward the final answer. Clearly, the more precise measurements should play a more important part in the calculation than the less precise values. To accomplish this we assign to the observations **weights** that are inversely proportional to the standard deviations of the observations. The derivations of the resulting equations can be found in the standard texts on statistics. We simply quote the results here.

Weighted Mean of a Set of Observations

Consider that we have a set of independently measured quantities, x_i, each of which has arisen from a statistical process that has yielded a value for its standard deviation, S_i. The **weighted mean** of the set of x values is defined to be

$$\bar{x} = \frac{\sum (x_i / S_i^2)}{\sum (1 / S_i^2)}$$

and the standard deviation of the weighted mean by

$$S^2 = \frac{\sum \left((x_i - \bar{x})^2 / S_i^2 \right)}{(N-1) \sum \left(1 / S_i^2 \right)}$$

Straight-Line Fitting by Weighted Least Squares

Consider that we have a set of values of a variable y measured as a function of x. As in Section 6–7 we assume that the x values are precise and that all the uncertainty is confined to the y values. We assume also that the y values are of unequal precision and have been assigned weights w. The equations by which we can calculate the slope m and the intercept b of the best line can be written as follows:

$$m = \frac{\sum w \sum wxy - \sum wx \sum wy}{\sum w \sum wx^2 - \left(\sum wx \right)^2}$$

$$b = \frac{\sum wy \sum wx^2 - \sum wx \sum wxy}{\sum w \sum wx^2 - \left(\sum wx \right)^2}$$

Because of the cumbersome nature of these equations, we have used abbreviated notation in which we have dropped the suffix i that should be attached to each of the quantities. Also, the term w_i that is used for the weight of each pair of measured values (x_i, y_i) is calculated in terms of the standard deviations of the y values as

$$w_i = \frac{1}{\left(S_{y_i} \right)^2}$$

The best estimates of the standard deviations for m and b can be written (as they were in Section 6–7) in terms of the deviations of the measured points from the best line. For a weighted least-squares fit, these deviations must now be weighted. The weighted value of S_y is given by

$$S_y = \sqrt{\frac{\sum w_i \delta_i^2}{N-2}}$$

The best estimates of the standard deviation for the slope and intercept can now be written as

$$S_m^2 = S_y^2 / W$$

and

$$S_b^2 = S_y^2 \left(\frac{1}{\sum w} + \frac{\bar{x}^2}{W} \right)$$

where

$$W = \sum \left(w(x - \bar{x})^2 \right)$$

and $\bar{x}$ is the weighted mean of the x values, given by

$$\bar{x} = \frac{\sum wx}{\sum w}$$

The suffix i has, as before, been omitted. Further detail on weighted least-squares fitting to straight lines is in the text by Squires that is listed in the Bibliography.

Appendix 3

Difference Tables
and the Calculus
of Finite Differences

A3–1 MATHEMATICAL FOUNDATIONS

The calculus of finite differences supplies a powerful tool for the treatment of measured variables. For the moment, however, let us consider the situation wholly from the mathematical point of view. After we have established mathematically the results we need, we can proceed to apply them to measurements.

Consider a known function $y = f(x)$ (see Figure A3–1) that can be expressed in terms of a Taylor expansion about the value $x = a$:

$$f(x) = f(a) + (x-a)\left(\frac{df}{dx}\right)_a + \frac{(x-a)^2}{2!}\left(\frac{d^2 f}{dx^2}\right)_a + \cdots$$

Such a function is said to be analytic at the point $x = a$. Any good book on calculus will provide more detail.

The function is defined along a continuous range of values on the scale of x, and, to make the theory applicable to measured variables, we must convert the formulation so that it refers to *discrete* values of x. Let these discrete

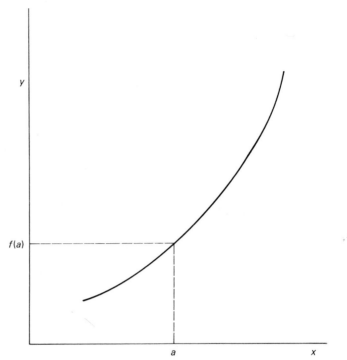

Figure A3–1 Graph of the function $y = f(x)$.

values of x be spaced equidistantly upward from $x = a$ at intervals of h, so that the values of x in which we are interested are

$$x = a, \quad x = a + h, \quad x = a + 2h, \quad x + 3h, \ \ldots$$

We can now calculate corresponding values of y for these discrete values of x. They are

$$f(a), \quad f(a + h), \quad f(a + 3h), \ \ldots$$

and we can illustrate these values on a graph as shown in Figure A3–2.

If we concentrate our attention on these discrete values of x and y, and if we wish to find a form of the Taylor expansion applicable to the discrete values, we can simulate the required derivatives by using finite differences. We define the quantity $\Delta f(a)$ to be

$$\Delta f(a) = f(a + h) - f(a)$$

Correspondingly, we also have,

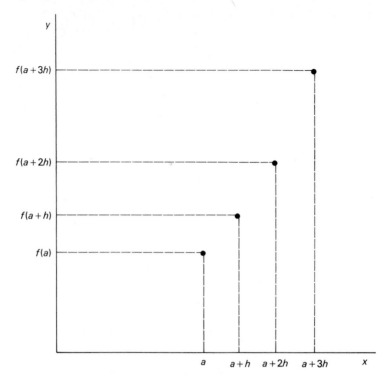

Figure A3–2 Values of $f(x)$ at values of x near $x = a$.

$$\Delta f(a+h) = f(a+2h) - f(a+h)$$

and so on, and these quantities are related to the first derivatives of the function at the various values of x. In a similar way we define the second differences

$$\Delta^2 f(a) = \Delta f(a+h) - \Delta f(a)$$

and so on for third and higher differences.

When we arrange all these differences beside the tabulated values of $f(x)$, we obtain a **difference table** for the values. A difference table for the function

$$y = 2x + x^3$$

is shown in Table A3–1.

TABLE A3–1 Difference Table for the Function $y = 2x = x^3$

x	y	Δ	Δ^2	Δ^3	Δ^4
1	3				
		9			
2	12		12		
		21		6	
3	33		18		0
		39		6	
4	72		24		0
		63		6	
5	135		30		0
		93		6	
6	228		36		0
		129		6	
7	357		42		0
		171		6	
8	528		48		0
		219		6	
9	747		54		
		273			
10	1020				

This table illustrates several important features of difference tables, including, for this example, the constancy of the third differences and the consequently zero value of the fourth differences.

Now let us consider the problem of obtaining values of y at values of x intermediate between the discrete values of x. Furthermore, let us find a way of calculating these intermediate values from the known values of y, instead of calculating them directly from the function itself. The advantage of such a procedure is that it will be available for later use on values for which we do not know the relevant function. To calculate these intermediate values, we must rewrite the Taylor expansion in a form that is compatible with the quantities found in the difference table and that also is suitable for the calculation of intermediate values. In Figure A3–3 the gradient of the function f at $x = a$ can be approximated by the ratio Δ/h. Corresponding values for the second derivative can be calculated in terms of the second difference Δ^2, and so on for higher derivatives.

Consider also a value of x intermediate between $x = a$ and $x = a + h$, and let it be specified by a new variable u that is defined by

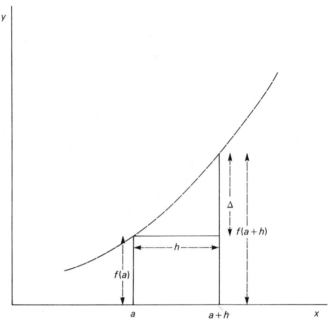

Figure A3–3 Approximation for the gradient of $f(x)$ at $x = a$.

$$x = a + uh$$

The Taylor expansion can now be rewritten in terms of the above-mentioned quantities to yield the intermediate values of y as

$$y = f(a) + u\Delta + \frac{1}{2!}u(u-1)\Delta^2 + \frac{1}{3!}u(u-1)(u-2)\Delta^2 + \ldots$$

This form of the Taylor expansion is known as the Gregory–Newton formula for interpolation. It can be used to calculate intermediate values whenever we have tabulated values of two variables. For example, such methods used to be commonly employed for interpolating in tabulated values of logarithms, trigonometric functions, and the like.

To use the Gregory–Newton formula, we construct a difference table as far as those differences that become either zero or else small enough that the error involved in the interpolation calculation becomes acceptably small. If we construct the table so that the value we are seeking lies between $x = a$ and $x = a + h$, the various differences we need for insertion in the formula will be found along the upper edge of the table. If we are seeking a value that lies between $x = a + h$ and $x = a + 2h$, the differences we need will be found along the next lower row, and so on.

Extrapolation can be carried out by a similar process. Suppose we have a set of values of y for values of x ranging from $x = a$ to $x = a + (n-1)h$. If we wish to calculate a value for y appropriate to $x = a + nh$, we must start with the supposition that the y values beyond $x = a + (n-1)h$ are determined by the same function as for the lower values. On the basis of that assumption, there is a simple method of finding y for $x = a + nh$; it is based on a process of extending the basic difference table. Starting with the column of differences that are constant, or sufficiently close to constancy for our purposes, we calculate successively the lower differences in terms of the higher until we reach the required value of y. The process is illustrated in Table A3–2, in which we have pretended that we did not know the function that y was of x, that we knew the values of y only for values of x up to 6, and that we desired the value of y for $x = 7$. In this way the table can be extended indefinitely to provide further values of the function as required.

TABLE A3–2 The Use of a Difference Table for Extrapolation

x	y	Δ	Δ^2	Δ^3
2	8			
		19		
3	27		18	
		37		6
4	64		24	
		61		6
5	125		30	
		91		
6	216			6
			30 + 6 = 36	
		91 + 36 = 127		
7	216 + 127 = 343			

The difference table for actual observations does not work out as neatly as did the two foregoing examples. For purposes of illustration these examples were constructed by using explicit mathematical functions, and absolute constancy was found at some level in the differences. Real observations differ from these examples in two respects. First, there is no guarantee, for a particular set of real observations, that a simple function that would lead to constancy of some difference column even exists. Second, even if some elementary function is appropriate, the presence of uncertainty in the observations makes it impossible to obtain complete constancy in any of the difference columns. We just use our best judgment in dealing with each situation as it arises. The situation is considered further at the end of this Appendix.

We can also use a difference table to construct a polynomial that will either reproduce the actual functional relationship between y and x (if such exists) or else provide an adequate approximation to a set of experimental observations. To do this we must rewrite the Gregory–Newton formula in a form suitable for our purpose. We earlier wrote it in terms of the variable u; we now wish to write it in terms of the variable x while still incorporating the differences Δ rather than the derivatives df/dx. Remembering the definition of u and using the equation

$$x = a + uh$$

we have

$$u = \frac{x - a}{h}$$

and the original form of the Gregory–Newton equation becomes

$$y = f(x) = f(a) + \frac{1}{h}(x - a)\Delta + \frac{1}{2!}\frac{1}{h^2}(x - a)(x - a - 1)\Delta^2$$

$$+ \frac{1}{3!}\frac{1}{h^3}(x - a)(x - a - 1)(x - a - 2)\Delta^3 + \ldots$$

The equation is now in the form we desire. If we insert in it the appropriate values of Δ, Δ^2, Δ^3, and so on, for a particular value such as $f(a)$, we generate the required polynomial in x.

As an example consider the difference table in Table A3–1 and choose the values in the top row. They are

$$f(a) = 3, \quad a = 1, \quad h = 1, \quad \Delta = 9, \quad \Delta^2 = 12, \quad \Delta^3 = 6, \quad \Delta^4 = 0$$

Inserting these values and performing some elementary algebra we obtain

$$y = 2x + x^3$$

Because this is the function we started with, we should not be surprised. However, we have confirmed the suitability of the Gregory–Newton formula for generating a polynomial that is consistent with a set of tabulated values. It is therefore of immense potential value for constructing a suitable polynomial when we are dealing with a set of measurements alone, with no idea of a suitable function to act as a model.

A3–2   APPLICATION OF DIFFERENCE TABLES
TO MEASURED VALUES

In the preceding section we illustrated the calculus of finite differences and difference tables by using mathematical functions. When we turn to measured variables and seek to apply these techniques, we encounter two differences: (1) We may not know a function that will provide an adequate fit to the observations, and (2) the variables will show scatter arising from uncertainty of measurement.

Consider first the case in which the measurements are precise, so that the scatter is negligible in comparison with the measured values. In this case the difference table will contain values that behave relatively regularly, and we can use it to perform immediately such procedures as interpolation and extrapolation. Furthermore, if a polynomial of a certain order will serve as a good fit to the observations, the differences of the appropriate order will turn out to be nearly constant, and the next differences will scatter around zero. We can then use the procedures of the preceding section to construct the appropriate polynomial.

If, on the other hand, our observations show larger scatter, we are faced with a somewhat different problem of interpretation. It is in principle possible to fit a polynomial exactly to any set of values, no matter how much scatter they show. In fact, to any set of values it is possible to fit an infinite number of polynomials, two of that infinity being represented by the curves in Figure A3–4.

So which polynomial do we want? Is it going to be one like that represented by the solid line in Figure A3–4? Under some circumstances this may be appropriate. On many other occasions, however, we may have good reason to believe that, measurement uncertainty apart, the basic behavior of the system is regular, and we really want the function corresponding to the dotted line in Figure A3–4. We have to consider, therefore, **smoothing** the observations, by which we mean choosing a function or curve that follows the observations in general terms but ignores deviations smaller than a selected limit. Many of the standard texts on measurement theory supply detailed descriptions of smoothing procedures. See, for example, the text by Whittaker and Robinson listed in the Bibliography.

It is not always clear how far we have to go in smoothing observations. We trade off simplicity of representation against possible loss of genuine detail in the behavior of the system. This takes good judgment on the part of the experimenter, and our decisions are not always greeted with universal

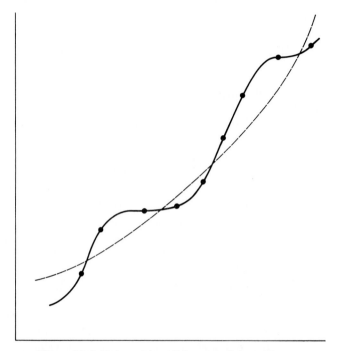

Figure A3–4 Fitting a polynomial to a set of observations.

approval. In any case, if we do want to make a choice of a certain order of polynomial to represent the observations, we can choose the corresponding difference column in the table to be constant and, on the basis of some aver-aged value of these differences, construct the polynomial we want.

If such a procedure is not to our liking and we are restricted to unavoidably noisy observations, our only alternative may be to use a least-squares procedure and thereby minimize the discrepancy between the observations and a function of a chosen type. Notice, however, the important distinction between the use of the difference table and the least-squares procedure. The difference table will tell us the coefficients of a polynomial that is implied by the observations; the least-squares procedure will simply give us the optimized parameters of a function whose general form we must choose for ourselves.

Appendix 4

Specimen Experiment

A4–1 EXPERIMENT DESIGN

System

We are given a spring suspended from a stand, a pan to hold weights attached to the lower end of the spring, a set of weights, and a stopwatch with a scale divided into fifths of a second.

Model

We are told that, on the assumption that the extension of a spring is proportional to the load on it, the period of oscillation, T, of a suspended load, m, given by

$$T = 2\pi\sqrt{\frac{m}{k}}$$

where k is a constant for a particular spring.

Requirement

We are asked to measure k for the spring with an uncertainty not greater than 10%.

Experiment Design

Following the steps listed in Section 5–3, we have the following:

1. The operating system consists of the spring alone. We have been given no information about the pan for the weights, or any way of weighing it, so we must proceed without that knowledge.

2. The model contains only two variables, load m and period of oscillation T, so it is simple to decide that our input variable, the one we can control, will be m and the output variable will be T.

3. To put the equation into straight-line form, our first idea might be to remove the square root. Squaring both sides of the equation gives us

$$T^2 = 4\pi^2 \frac{m}{k}$$

Comparison with the straight-line equation

vertical variable = slope × horizontal variable

suggests that we could choose

vertical variable = T^2

horizontal variable = m

$$\text{slope} = \frac{4\pi^2}{k}$$

This is an acceptable choice, but the unknown, k, appears in the denominator of the slope. To simplify later arithmetic, it is equally valid and more convenient to write the equation

$$m = \frac{k}{4\pi^2} T^2$$

where

vertical variable = m

horizontal variable = T^2

$$\text{slope} = \frac{k}{4\pi^2}$$

4. The range of the input variable, m, may be governed by the weights we have been given. In addition, we should consider the possibility of over-

loading the spring. Has anyone suggested, or is it written anywhere, that loads should be restricted? We might try a few weights on the pan to see how the spring behaves. One way or another, we can choose a range of m that we feel comfortable with. The range of T values presents no problem because it is determined by the system.

5. Let us suppose that our weights are sufficiently precise so that their uncertainty need not be considered. They are not totally precise, of course, and if we want to know what uncertainty they do have, we look in the supplier's catalogue.

The only uncertainty, therefore, arises from the timing measurements, and that uncertainty depends on the precision with which we are capable of timing the oscillations. The only way to find that out is to try it. We choose a typical load, start the oscillations, and measure the time interval for, say, 10 oscillations. We must now decide what determines the uncertainty of the measurement. Is it the accuracy of reading the stopwatch, or is it our ability to watch the oscillations and start or stop the watch appropriately? Obviously, we must test this by trying the measurement again, and we must continue to probe the measuring system in this way until we are sure we know our capabilities. We may decide, as in the present case, that we are sure that we can measure time intervals with an uncertainty of ±0.3 s.

This, however, does not complete our consideration of precision; we must evaluate the effect on our k values of this uncertainty in T. It is difficult to plan ahead exactly, because we shall obtain our final value of k from the graph, but it is only prudent to check that our individual measurements have adequate precision to contribute significantly to the final result.

For example, suppose we time a certain number of oscillations that give a time interval, t, of 2 seconds. What would be the contribution of our ±0.3 s to the uncertainty in k? k is a function of t^2. Therefore

$$\text{RU}(k) = 2 \times \frac{0.3}{2} = 30\%$$

Clearly, such a measurement makes little significant contribution to the determination of k with 10% precision. If we want 10% precision in k, we need 5% precision in t, and that requirement imposes limits on our measuring process for t. We can determine these limits as follows:

If

$$\text{RU}(t) \text{ must not be greater than } 5\%$$

that is,

$$\frac{0.3}{t} < 0.05$$

then

$$t > \frac{0.3}{0.05} = 6 \text{ s}$$

Thus, whatever the actual value of T, if we time a number of oscillations that gives a total time interval of 6 seconds or more, the uncertainty in our timing measurements is likely to be acceptable. For convenience we might choose a constant number of oscillations for the various loads, but if we were short of time in a long experiment, we could choose for each load the number of oscillations that gives us a satisfactory value of t.

6. Let us decide, as a first guess, that we shall measure the time for 10 oscillations, knowing that even for the lowest load this gives us a measured time interval in excess of 6 s, and that we shall time oscillations for loads of 0.1, 0.15, 0.2, 0.25, 0.3, 0.35, 0.4, 0.45, and 0.5 kg. Since we want to plot m versus T^2 and incorporate the value of the absolute uncertainty in T^2, that is, $2T \times \text{AU}(T)$, we should lay out the table that expresses our measurement program to have the following headings:

Load, m, kg	Number of Oscillations	Time, t, s	Period, T, s	Period^2, T^2, s^2	AU(T^2), s^2

Experimental Results

The next step is to make the actual measurements and fill in the table with the measured values of t and the values calculated for Period and for Period^2 with its absolute uncertainty. The result of this process will then appear as in the following table.

Load, m, kg	Number of Oscillations	Time, t, s	Period, T, s	Period^2, T^2, s^2	AU(T^2), s^2
0.10	10	8.2 ± 0.3	0.82 ± 0.03	0.67	± 0.05
0.15	10	9.8	0.98	0.96	0.06
0.20	10	10.7	1.07	1.14	0.06
0.25	10	11.5	1.15	1.32	0.07
0.30	10	12.5	1.25	1.56	0.08
0.35	10	13.0	1.30	1.69	0.08
0.40	10	13.8	1.38	1.90	0.08
0.45	10	14.5	1.45	2.10	0.09
0.50	10	15.2	1.52	2.31	0.09

These values of m and T^2 must now be graphed. Each plotted value must contain its range of uncertainty. m has no uncertainty and T^2 is uncertain by the amount listed in the final column, so that each point on the graph will be a little horizontal line. Once the values are plotted, the graph will look as shown in Figure A4–1.

The next step is to interpret what we see in terms of the categories described in Section 6–4. We first observe the scatter of the points and consider if it is consistent with our prior estimate of the uncertainty. In the present case there seems to be reasonable consistency between the scatter and the estimated uncertainty, and no further consideration of this point seems to be required at this stage. The next point to consider is the extent to which the

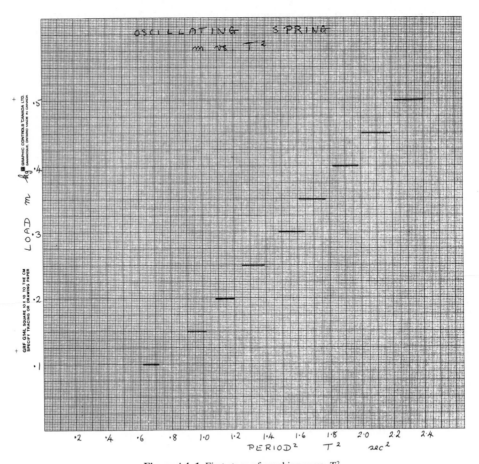

Figure A4–1 First stage of graphing m vs. T^2.

behavior of the system is consistent with the model. The model predicted a straight line passing through the origin, and we must judge our graph against that. We can see immediately that in our case the correspondence with straight-line behavior seems quite adequate over the whole range. We are justified, therefore, in including all the points when we decide on our choice of lines.

With regard to intercept, however, the situation is different. A glance at the graph makes it clear that we are going to have an intercept on the T^2 axis that cannot be ascribed to measurement uncertainty. We return to consider this discrepancy later, but in the meantime we can note that the final value of k will be obtained from the slope alone, and that the slope can be calculated even in the presence of an unexpected intercept.

The next step is to draw lines so that we obtain values for the slopes. One choice is to draw our "best" line by eye and also lines that represent the maximum and minimum slopes permitted by the range of uncertainty in the scatter. At this stage, the graph will look as shown in Figure A4–2.

We now have to read values off the graph that will enable us to calculate these slopes. For each line we look for convenient intersections with the graph paper, illustrated in Figure A4–3, that give us the coordinates of points at the top and bottom of the line.

On the present graph, the chosen intersections are indicated by arrows, and the appropriate coordinates are marked. Given these coordinates, it is easy to calculate the slopes.

Steepest line:

$$\text{slope} = \frac{0.55 - 0.075}{2.40 - 0.73}$$
$$= 0.284 \text{ kg s}^{-2}$$

Central line:

$$\text{slope} = \frac{0.55 - 0.05}{2.50 - 0.52}$$
$$= 0.253 \text{ kg s}^{-2}$$

Least steep line:

$$\text{slope} = \frac{0.55 - 0.075}{2.64 - 0.51}$$
$$= 0.223 \text{ kg s}^{-2}$$

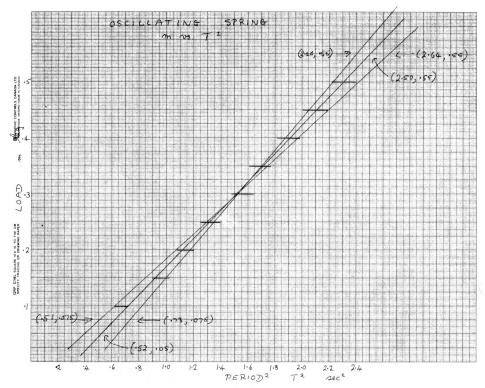

Figure A4–2 Final form of graph of m versus T^2.

The corresponding values of k can now be calculated by using

$$\text{slope} = \frac{k}{4\pi^2}$$

which gives

$$k = 4\pi^2 \times \text{slope}$$

Highest value:

$$k = 11.211 \text{ kg s}^{-2}$$

Middle value:

$$k = 9.988 \text{ kg s}^{-2}$$

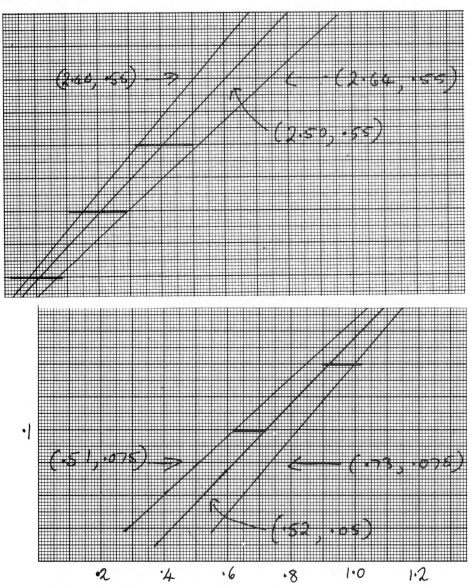

Figure A4–3 The identification of intersections at the top and at the bottom of the graph to permit calculation of the slopes.

Lowest value:

$$k = 8.804 \text{ kg s}^{-2}$$

Now that we have a measure of the overall uncertainty of the k values, we can round off the values to obtain our final statement about k and its uncertainty:

$$k = 10.0 \pm 1.2 \text{ kg s}^{-2}$$

The final figure for the uncertainty at 12% of the k value is slightly greater than the target figure of 10%, but perhaps we can say that we have come close enough to our objective. If any further reconsideration is forced on us, we could go back to the beginning and reassess our basic measurement uncertainty in timing. Certainly, the low scatter of the points in comparison with the estimated uncertainty in the upper part of the graph suggests that we were slightly pessimistic about our estimate of ±0.3 s in timing, and reappraisal might allow us to refine that estimate.

Given the completion of our calculations for k, we now have to return to the question of the unexpected intercept. We satisfied ourselves that it was harmless because our k value was obtained from the slope, and that could be determined accurately even in the presence of the intercept. Nevertheless, we should not ignore it altogether, because it constitutes failure of correspondence between the model and the system, and it is not good practice in experimenting to leave things like that unconsidered. In guessing at possible sources of the discrepancy, we note that it seems to be associated with some load not counted in the m values, for when our added load, m, is zero, the graph tells us that we would still observe a finite frequency of oscillation. What could give rise to such uncounted mass? One obvious guess would be the pan on which the weights were placed. Another obvious guess would be the mass of the spring itself. Without further investigation, we cannot be certain that either of these guesses is good, but our explanation for the unexpected intercept seems reasonable enough so that we are probably justified in terminating our present experiment at this point and leaving confirmation of our guesses to further experimenting.

A4–2   REPORT

In this section we give a version of a final report that could be written on the basis of the experiment that we have just described. The report is written according to the suggestions offered in Chapter 7. Only the final version of the report is given; the details of its construction and their correspondence with the suggestions in Chapter 7 can be elucidated by comparing the report with the various sections of Chapter 7.

MEASUREMENT OF A SPRING CONSTANT BY AN OSCILLATION METHOD

Introduction

The stiffness of a spring can be measured by timing the oscillation of a suspended load. For an elastic spring (extension $\propto$ load) it can be proved that the period of oscillation, T, of a suspended mass, m, is given by

$$T = 2\pi\sqrt{\frac{m}{k}} \tag{1}$$

where k is a constant for a particular spring. The objective of the present experiment is to measure the value of k for a spring with an uncertainty not greater than 10%.

Equation (1) can be rewritten to read

$$m = \frac{k}{4\pi^2}T^2$$

which is linear in m and T^2 with slope $k/4\pi^2$. Thus, by measuring the variation of oscillation period with load we shall be able to plot a graph of m versus T^2 and obtain the value of k from the slope.

Procedure

Using the apparatus shown in Figure 1, we measured the variation of oscillation period with load.

The loads consisted of a Cenco weight set ranging from 0.1 kg to 0.5 kg with stated precision 0.04%.

The period of oscillation was measured by timing a number of oscillations using a stopwatch graduated in fifths of a second.

Results

The measurements of load and oscillation period are shown in Table 1.

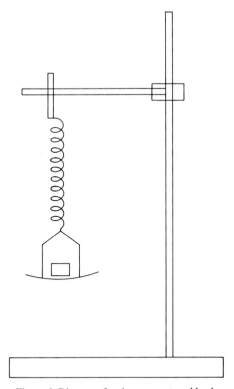

Figure 1 Diagram of spring, support, and load.

TABLE 1 Variation of Oscillation Period with Load

Load, m, kg	Number of Oscillations	Time, t, s	Period, T, s	Period^2, T^2, s^2
0.10	10	8.2 ± 0.3	0.82	0.67 ± 0.05
0.15	10	9.8	0.98	0.96 ± 0.06
0.20	10	10.7	1.07	1.14 ± 0.06
0.25	10	11.5	1.15	1.32 ± 0.07
0.30	10	12.5	1.25	1.56 ± 0.08
0.35	10	13.0	1.30	1.69 ± 0.08
0.40	10	13.8	1.38	1.90 ± 0.08
0.45	10	14.5	1.45	2.10 ± 0.09
0.50	10	15.2	1.52	2.31 ± 0.09

The uncertainty shown for the measured times was estimated by visual judgment to be the outer limits for possible values of time, and so the calculated values for the uncertainty of T^2 also represent outer limits of possibility.

The graph of the m and T^2 values is shown in Figure 2.

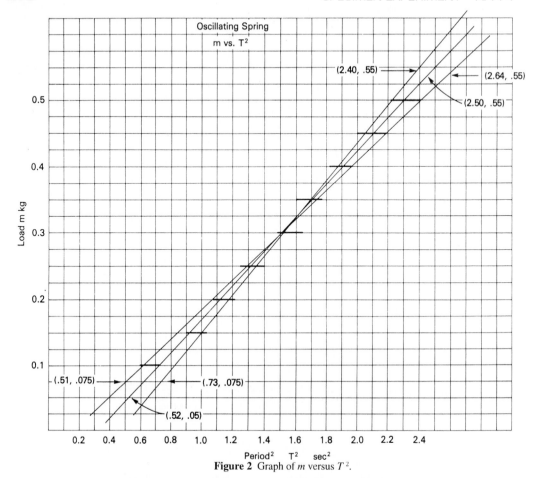

Figure 2 Graph of m versus T^2.

The value of k and its uncertainty was calculated from the slopes of the three lines shown, giving

$$k = 10.0 \pm 1.2 \text{ kg s}^{-2}$$

Discussion

Using an oscillation method, we have measured the value of the stiffness constant for a spring to be 10.0 ± 1.2 kg s^{-2}. The model represented by equation (1) predicted for the m versus T^2 graph a straight line passing through the origin. In our experiment, the variation of m with T^2 proved to be consistent with a straight line within the present limits of uncertainty. Instead of passing through the origin, however, the lines shown in Figure 2 can be seen to have a finite intercept on the T^2 axis that cannot be ascribed to measurement uncertainty. Our value of k, however, was obtained from the slope alone and should be free from error arising from factors that would give rise to an intercept.

Because the intercept gives a finite value for T at $m = 0$, it seems to be associated with the presence of some load not included in the measured values of m. Although we have not tested these possibilities, we can suggest that such extra load could arise from the pan supporting the weights and also from the mass of the spring itself.

Bibliography

The subject of experimenting is vast, and the literature is correspondingly extensive and diverse. To meet all our varied needs there is no satisfactory substitute for going to the library and becoming familiar with its range of offerings. The following list contains some relatively recent books and some classics, but it is very far from exhaustive. It is intended only to provide a few suggestions that may serve as a starting point for individual study.

BACON, R. H. "The 'Best' Straight Line Among the Points," *American Journal of Physics,* **21**, 428 (1953).

BARFORD, N. C. *Experimental Measurements: Precision, Error and Truth.* Addison-Wesley, 1967.

BARRY, R. A. *Errors in Practical Measurement in Science, Engineering and Technology.* Wiley, 1978.

BEERS, Y. *Introduction to the Theory of Error.* Addison-Wesley, 1957.

BEVINGTON, P. R. *Data Reduction and Error Analysis for the Physical Sciences.* McGraw-Hill, 1969.

BRADDICK, H. J. J. *The Physics of the Experimental Method.* Chapman and Hall, 1956.

BRAGG, G. M. *Principles of Experimentation and Measurement.* Prentice-Hall, 1974.

BRINKWORTH, B. J. *An Introduction to Experimentation.* The English University Press, 1973.

COHEN, I. B. *Revolution in Science.* Harvard University Press, 1985.

COOK, N. H., RABINOWICZ, E. *Physical Measurement and Analysis.* Addison-Wesley, 1963.

COX, D. R. *Planning of Experiments.* Wiley, 1958.

DEMING, W. E. *Statistical Adjustment of Data.* Wiley, 1944.

DRAPER, N. R., SMITH, H. *Applied Regression Analysis.* Wiley, 1981.

FRETTER, W. B. *Introduction to Experimental Physics.* Prentice-Hall, 1954.

FREUND, J. E. *Modern Elementary Statistics.* Prentice-Hall, 1961.

HALL, C. W. *Errors in Experimentation*. Matrix Publishers, 1977.

HARRÉ, R. *Great Scientific Experiments*. Oxford University Press, 1983.

HORNBECK, R. W. *Numerical Methods*. Quantum Publishers, 1975.

JEFFREYS, H. *Scientific Inference*. Cambridge University Press, 1957.

KUHN, T. S. *The Structure of Scientific Revolutions*. University of Chicago Press, 1970.

LEAVER, R. H., THOMAS, T. R. *Analysis and Presentation of Experimental Results*. Macmillan, 1974.

LINDLEY, D. V., MILLER, J. C. P. *Cambridge Elementary Statistical Tables*. Cambridge University Press, 1958.

LYONS, L. A Practical Guide to Data Analysis for Physical Science Students. Cambridge University Press, 1991

MARGENAU, H., MURPHY, G. M. *The Mathematics of Physics and Chemistry*. Van Nostrand, 1947.

MENZEL, D. H., JONES, H. M., BOYD, L. G. *Writing a Technical Paper*. McGraw-Hill, 1961.

PARRATT, L. G. *Probabiliiy and Experimental Errors in Science*. Wiley, 1961.

RABINOVICH, S. *Measurement Errors, Theory and Practice*. American Institute of Physics, 1993.

RABINOWICZ, E. *An Introduction to Experimentation*. Addison-Wesley, 1970.

ROSSINI, F. D. *Fundamental Measures and Constants for Science and Technology*. CRC Press, 1974.

SCHENCK, H. *Theories of Engineering Experimentation*. McGraw-Hill, 1961.

SHAMOS, M. H. (editor). *Great Experiments in Physics*. Holt-Dryden, 1960.

SHCHIGOLEV, B. M. *Mathematical Analysis of Observations*. American Elsevier, 1965.

SQUIRES, G. L. *Practical Physics*. McGraw-Hill, 1968.

STANTON, R. G. *Numerical Methods for Science and Engineering*. Prentice-Hall, 1961.

STRONG, J. *Procedures in Experimental Physics*. Prentice-Hall, 1938.

STRUNK, W., WHITE, E. B. *The Elements of Style*. Macmillan, 1979.

TAYLOR, J. R. *An Introduction to Error Analysis*. University Science Books, 1982.

TOPPING, J. *Errors of Observation and Their Treatment*. The Institute of Physics, London, 1955.

TUTTLE, L., SATTERLEY, J. *The Theory of Measurements*. Longmans Green, 1925.

WHITTAKER, E. T., ROBINSON, G. *The Calculus of Observations*. Blackie, 1944.

WILSON, E. B. *An Introduction to Scientific Research*. McGraw-Hill, 1952.

WORTHING, A. G., GEFFNER, J. *Treatment of Experimental Data*. Wiley, 1946.

Answers to Problems

CHAPTER 2:

1. 142.45 ± 0.15 cm; 0.11%
2. 1.245 ± 0.005 A,
   3.3 ± 0.1V; 0.4%, 3.0%
3. 0.5 minute
4. (a) 10 cm, (b) 2 cm
5. 7.7%
6. 0.6%
7. 26 cm^2

8. 0.30 ohm
9. 9.77 ± 0.04 m s^{-2}, 0.4%
10. 1800 kg m^{-3}
11. 0.110 m, 0.0012 m, 1.08%
12. 0.8 nm, 0.24%
13. 14.3 ± 0.1; 14.25 ± 0.15
14. 6.75 ± 0.03

CHAPTER 3:

1.

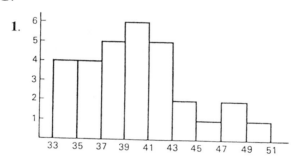

2. Between 39 and 41, 38
3. 38.3
4. 4.39

5. 0.80
6. 0.58
7. (a) 33.9 to 42.7,
   (b) 29.5 to 47. 1
8. (a) 37.5 to 39. 1,
   (b) 36.7 to 39.9
9. (a) 3.8 to 5.0,
   (b) 3.2 to 5.5

10. 0.16
11. Rejection
12. Sample 34, 47, 43, 40, 32 has mean
    39.2 and standard deviation 5.56
    Sample 36, 40, 38, 43, 34 has mean
    38.2 and standard deviation 3.12
13. More than 130
14. More than 200
15. 0.47 mm^2
16. 0.21×10^{-9} m
17. 0.11 m s^{-2}

CHAPTER 5:

1. No

2. range $\propto \left(\dfrac{\text{velocity}}{g} \right)^2$

3. pressure $\propto \dfrac{\text{surface tension}}{\text{radius}}$

4. period $\propto \sqrt{\dfrac{\text{moment of inertia}}{\text{rigidity constant}}}$

5. deflection $\propto \left(\dfrac{\text{load force}}{y \times \text{radius}^2} \right)^a$

$\times \left(\dfrac{\text{length}}{\text{radius}} \right)^b \times \text{radius}$

where a and b are arbitrary constants

6. s vertically, t^2 horizontally, slope is $a/2$

7. T vertically, $f^2 \ell^2$ horizontally, slope is $4m$

8. P vertically, v^2 horizontally, slope is $\rho / 2$

9. T^2 vertically, $\cos \alpha$ horizontally, slope is $4\pi^2 \ell / g$

10. d vertically, $W\ell^3$ horizontally, slope is $4 / Yab^3$

11. h vertically, $1/R$ horizontally, slope is $2\sigma / \rho g$

12. p vertically, T horizontally, slope is R/v

13. fv_0 vertically, $f - f_0$ horizontally, slope is v

14. ℓ vertically, ΔT horizontally, slope is $\ell_0 \alpha$ and intercept is ℓ_0, whence α

15. $\sin \theta_1$ vertically, $\sin \theta_2$ horizontally, slope is n_2 / n_1

16. $1/s$ vertically, $1 / s'$ horizontally, each intercept is $1/f$; or ss' vertically, $s + s'$ horizontally, slope is f

17. $1/C$ vertically, ω^2 horizontally, slope is L

18. F vertically, $1 / r^2$ horizontally, check for linearity

19. F vertically, $i_1 i_2 / r^2$ horizontally, check for linearity; also check F vs. i_1, F vs. i_2 and F vs. $1 / r^2$ separately, while holding other variables constant

20. $\log_e Q$ vertically, t horizontally, slope is $-1/RC$

21. Z^2 vertically, $1 / \omega^2$ horizontally, slope is $1 / C^2$ and intercept is R^2

22. m^2 vertically, $m^2 v^2$ horizontally, slope is $1 / c^2$ and intercept is m_0^2

23. $1 / \lambda$ vertically, $1 / n^2$ horizontally, slope is $-R$, intercept is $R/4$

CHAPTER 6:

1. (c) $0.00129 \text{ ohm}^2\text{s}^2$
   (d) 0.00572 henry
   (e) $0.00145 \text{ ohm}^2\text{s}^2$,
       $0.00117 \text{ ohm}^2\text{s}^2$
   (f) The measured value of L
       can range from 0.00544
       henry to 0.00606 henry
   (g) 6.00 ohm
   (h) The measured value of R
       can range from 5.39 ohm
       to 6.43 ohm.
   (i) $L = 0.0057 \pm 0.0004$ henry
       $R = 6.0 \pm 0.6$ ohm
2. Mean is 17.54 and the standard
   deviation of the mean is 0.26

3. (a) Slope is $0.0499 \text{ ohm deg}^{-1}$
       and the intercept is 11.92 ohm
   (b) $\alpha = 0.00419 \text{ deg}^{-1}$
   (c) $0.0024 \text{ ohm deg}^{-1}$
       0.16 ohm
   (d) 0.00021 deg^{-1}
   (e) $\alpha = 0.00419 \pm 0.00021 \text{ deg}^{-1}$
       $R_0 = 11.92 \pm 0.16$ ohm
4. $a = 2.12$, $b = 2.98$
5. (a) $i = 0.5e2^v$
   (b) $y = 0.6x^{2.4}$
   (c) $f = 6.2e^{-365/T}$
6. (b) 0.73
7. 0.99

Index